Society of Earth Scientists Series

Series editor

Satish C. Tripathi, Lucknow, India

The Society of Earth Scientists Series aims to publish selected conference proceedings, monographs, edited topical books/text books by leading scientists and experts in the field of geophysics, geology, atmospheric and environmental science, meteorology and oceanography as Special Publications of The Society of Earth Scientists. The objective is to highlight recent multidisciplinary scientific research and to strengthen the scientific literature related to Earth Sciences. Quality scientific contributions from all across the Globe are invited for publication under this series.

More information about this series at http://www.springer.com/series/8785

K.S. Valdiya

Prehistoric River Saraswati, Western India

Geological Appraisal and Social Aspects

The Society of Earth Scientists
2008

Springer

K.S. Valdiya
Geodynamics Unit
Jawaharlal Nehru Centre for Advanced
Scientific Research
Bengaluru
India

ISSN 2194-9204 ISSN 2194-9212 (electronic)
Society of Earth Scientists Series
ISBN 978-3-319-44223-5 ISBN 978-3-319-44224-2 (eBook)
DOI 10.1007/978-3-319-44224-2

Library of Congress Control Number: 2016949590

Printed on acid-free paper

This Springer imprint is published by Springer Nature
The registered company is Springer International Publishing AG
The registered company address is: Gewerbestrasse 11, 6330 Cham, Switzerland

Series Editor Foreword

Rivers are sites for origin and diversity of human civilisations and any change in their course/source of water has a great impact on cultural development. Many a great civilisation has vanished in history due to the changes in the river systems. Integrated study of river system is an important domain for earth scientists, and lost rivers remain a subject of debate and contention. The existence of lost prehistoric River Saraswati, one of the sacred rivers of India along which great Harappan Civilization developed, is also a topic of debate within the scientific community. Geoscientific studies have been carried out extensively by different workers on various aspects. Professor K.S. Valdiya, a noted teacher, scholar and scientist, is associated with the geological studies of River Saraswati since long and has also drawn excellent scientific conclusions from ancient Indian literature and archaeological studies. Thus, this monograph, apart from its scientific value, has removed the dividing line between geology and archaeology. It also suggests that even some ruined cultural site may pose a question to geologists and may initiate geological studies to understand natural reasons/disasters, if any, associated with the destruction. It is because that the book is expected to be very useful and interesting for both earth scientists and archaeologists.

Satish C. Tripathi

Preface

Students of geology have long been reading graphic descriptions of the "lost" River Saraswatī in Edwin Pascoe's *A Manual of Geology of India and Burma*, Volume One, published in 1950 by the Geological Survey of India. The article I wrote in the popular Indian magazine *Dharmayug* in 1968 on the robbing of the River Saraswatī water by the Gangā evoked widespread interest and curiosity. In 1979, 11 years later, in well-researched papers Bimal Ghose and associates at Central Arid Zone Research Institute, Jodhpur followed by Yashpal et al. (1980) of Space Application Centre, Ahmadabad (Indian Space Research Organization) charting the course of the river on the basis of satellite imagery provided a strong scientific support to the postulation that once a great river flowed through the now Haryānā and the desolate and dreary land of the Thār Desert. My 1998 article in a science journal *Resonance* generated keen interest in the community of scientists. Among them was Prof. Roddam Narasimha, FRS, who invited me to give a talk at the National Institute of Advanced Studies to a large gathering of scholars and laymen and suggested that I write a book on the Saraswatī. In the audience was the late Prof. Satish Dhawan, former Chairman of Indian Space Research Organization (ISRO) and ex-Director of Indian Institute of Science. He persuasively urged me to write a booklet on the Saraswatī and made the ISRO to subsidize its publication in 2002.

The book was *not* a best seller, but within a few years not a single copy was left in the publisher's stock. I continued to get requests to spare my own copies— even from those who wanted to make a film or a documentary on the river that nourished for over three thousand years the vibrant Harappan Civilization. I do not know what happened after I sent my little book to seven or eight film persons.

In recent years, one or the other of the quite many scientists has come out with the comprehensive studies on geomorphological, sedimentological, geochronological, hydrological and remote-sensing aspects of the Saraswatī River. A majority of earth scientists have come to the conclusion that it was a large river that had abundant discharge and brought voluminous quantity of sediments from the Himālayan province. And there are quite a few who deduced that the Saraswatī was a monsoonal rain-fed river originating in the Outer Siwālik and its foothills. Studying in great detail the pattern of human settlements of the Harappan

Civilization located on the banks of an extraordinarily wide, sand-filled nearly waterless water course known as Ghagghar in Haryānā, Hākrā in Cholistān and Nārā in Sindh, the archaeologists harboured no doubt on this water course being the legendary Saraswatī full of life and bounty.

I felt the urge to write again on the geological aspects of the river that was the lifeline of the people of the progressive and vibrant society that chose to cling to the bank of this river for thousands of years, and lived a buoyant life full of appreciation for arts, crafts, commerce, agriculture and nature. Enormous evidence that archaeologists have gathered points to the Harappan Civilization being nourished by the life-sustaining Saraswatī.

The invocation of rivers Gangā, Yamunā, Sindhu, Saraswatī, Godāvarī, Narmadā and Kāverī in all ceremonies of the people of the largest segment of the Indian society indicates the exalted position the rivers occupy in our scheme of things. The *Rigved* verse 1:3:12, extolling the Saraswatī as a purifier endowed with riches and treasures of intellect and enlightenment demonstrates how great the Saraswatī River was to the people in the Vedic times. The accounts given in the ancient Indian literature, such as *Rāmāyan*, *Purāns* and *Mahābhārat*, cannot be rubbished and ignored, for they do contain grains of truths, the kernels of revealing facts. Keeping in mind the perspective of the geology, evolutionary history of the Indian subcontinent and the geomorphological layout of NW India, if one reads the texts of the *Purāns* and the epics, it would be clear that the geographical descriptions of mountains, rivers and landforms in the works of noted ancient writers Vālmiki and Krishna Dwaipāyan 'Vyās' are quite accurate in descriptions. Undoubtedly, descriptions and narratives are heavily enmeshed in verbose language, are replete with metaphors, and are embellished with allegories. Shorn of these superfluities, the *shlokās* (verses) do provide material of historical value. One can find geological reality lying hidden in the narratives if the texts are read without metaphors and superfluous phrases. The reality that emerges from the narratives in the *Purāns* and the epic *Mahābhārat* is that there was a great river which started drying up during the *Purān* times and was practically waterless by the time the *Mahābhārat* was written by Krishna Dwaipāyan 'Vyās' sometime after 3500 years Before Present.

In writing on the legendary Saraswatī—it may be emphatically stated—I was not swayed by my Purānic sympathies. Rather, I viewed the scenario within the framework of geological parameters and rigorously evaluated all inferences and surmises on the anvil of the principles of geodynamics. Presented in this book is the geological history of river that is now represented by an extraordinarily wide and waterless channel snaking through the vast floodplain in northwestern Haryānā and adjoining Rājasthān, by the considerably thick and extensive riverine deposits containing material of Himālayan parentage, and testified by the thousands of years old freshwater lying concealed in underground reservoirs in the heart of sandy Thār Desert, by the dense clusters of ruins of human settlements on the banks of the dry water courses, and by the occurrence of an ancient seaport opposite the remnants of a delta in what is today a salt-encrusted marshy flat of the Rann of Kachchh, the seaport speaking eloquently of navigable river that discharged into the Arabian Sea.

It was the river that vanished as a consequence of tectonic upheaval in the foothills of the Himālaya.

I wish to emphasize that this is not a scholarly treatise, nor a comprehensive analysis. It is just a geologist's interpretation of the mass of facts of varied kind, presented with a modest objective of providing a few credible examples that testify to the existence of a Himālayan-born river that in the prehistoric time was the lifeline of the people who had settlements in the land between the well-watered floodplains of the Sindhu and the Gangā River systems.

Despite a section of historians dismissing it as a figment of the imagination, as a fantasy, I believe that the Saraswatī was not a fantasy. It was a reality. This modest work endeavours to portray that reality in the land with many layers of history.

Bengaluru, India K.S. Valdiya

Acknowledgement

Professor Kanchan Pande at the Indian Institute of Technology, Mumbai, not only spurred me to write this book but also helped me in many ways, including the redrafting of diagrams and critically reviewing the manuscript. Dr. Satish C. Tripathi, at the Society of Earth Scientists, Lucknow encouraged me to write without loss of time and offered to publish under his society's banner.

I am profoundly grateful to Dr. Ravindra S. Bisht, formerly Joint Director-General of the Archaeological Survey of India (New Delhi), Dr. Amal Kar, formerly Principal Scientist and Head of Natural Resources and Environment Division, Central Arid Zone Research Institute (Jodhpur), Dr. Anil Kumar Gupta, formerly Senior Scientist at the Regional Remote Sensing Centre (Jodhpur) of Indian Space Research Organization, Dr. B.K. Bhadra Scientist/Engineer S.F. at Regional Remote Sensing Centre (ISRO), and Prof. Jyotiranjan S. Ray at Physical Research Laboratory (Ahmadabad) for rigorously reviewing the manuscript and giving very valuable suggestions for improvement.

Dr. Ravindra S. Bisht spared many photographs on the Harappan Civilization. The photographs of the Himālayan terrain were taken by Shree Anup Sah and Shree Niraj Pant. I am extremely grateful to them. I thank Ms. Sapna Shinde for drawing the line illustration and Dr. Jaishri Sanwal for helping in many ways.

K.S. Valdiya

Contents

Chapter 1
Mighty Tempestuous River

Land Without Rivers

A look at the map of the Indo-Gangetic Plains would make it quite obvious that a vast expanse of the land between the Yamunā and Satluj rivers is a parched realm of sandy plain bereft of the bounty of rivers (Fig. 1.1). In this riverless land the annual rainfall is no more than 15–50 cm. Compounding the problem of pervasive aridity, droughts visit the region at an interval of 2–5 years. While the central, eastern and the western sectors of the Indo-Gangetic Plains are watered profusely round the year by multitude of Himālayan rivers, the great expanse of the land encompassing Haryānā, the northern and western parts of Rājasthān, and the adjoining region of Pākistān is a riverless country (Fig. 1.2).

Despite having no rivers, except for a few ephemeral streams, this parched domain contains evidence of living of primitive people of the Stone Ages. In this very land beginning around 5500 yr B.P., the Harappa Civilization blossomed 4600–3300 years ago (Kenoyer 1998; Rao 1999; Bhan 1972) (Fig. 1.2). And it was this part of India that witnessed epochal events and stirring developments of history, not only in the historical period but also in the Purān times more than 3500 years ago (Valdiya 2012). In spite of aridity, caravan after caravan of traders, invaders and conquerors traversed the land. Among the invaders were Alexander the Great (327 B.C.E), Mahmūd Ghazanavi (A.D. 997–1025), Mohammad Ghori and Ahmad Shāh Abdāli (A.D. 1760) who took the road across this riverless region experiencing the tyranny of aridity. But the famous traveller Ibn Batūtā (A.D. 1325–1354) saw greenery of sugarcane and paddy crops in the fluvial reaches of the Ghagghar. There were such bustling trading towns as Mārot, Fort Abbās, Sirsā, Bhadrā and Hānsi along the road that traversed the dry land.

Even more surprising is the fact that the desert tracts continue to support a burgeoning population in the present: the Thār is described as world's most crowded desert where the population between 1909 and 1971 grew at the rate of 158 %, compared to less than 132 % in the rest of the country in the same period

K.S. Valdiya, *Prehistoric River Saraswati, Western India*,
Society of Earth Scientists Series, DOI 10.1007/978-3-319-44224-2_1

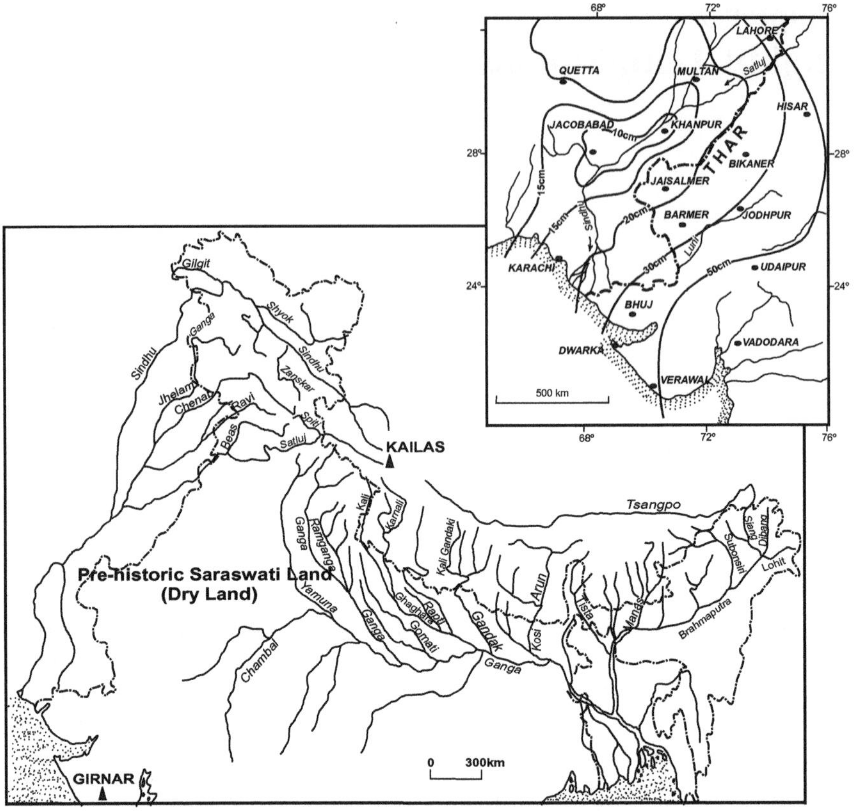

Fig. 1.1 While the central-eastern and the western sectors of the Indo-Gangetic Plains are profusely watered by Himālayan rivers, the vast expanse between the rivers Yamunā and Satluj is bereft of the bounty of perennial rivers. *Inset*; West of the Arāvali Ranges and east of the Sindhu floodplain the vast swathe of semi-arid to arid land receives annual rainfall no more than 15–50 cm (Valdiya 2002)

(Gupta and Prakash 1975). In 2011 the human population density was 133 persons/km^2 (Kar 2014b). Even the livestock population grew rapidly from 72/ha in 1951–175/ha in 1971 (Chaudhary et al. 2011). While the livestock population was 25.52 million in 1951, it was recorded much higher at 49.14 million in 2003. And the livestock population density in 2011 was 115/km^2 (Kar 2014b). Obviously, the desert sediments are rich in nutrients—the materials that the flowing rivers bring from the mountains made of varied assemblages of rocks. Indeed the vast expanse of the land without rivers at present is covered by thick columns of sediments brought by past rivers and deposited in their channels and floodplains. On the top, it is mostly aeolian sands emplaced by the blowing winds of the desert.

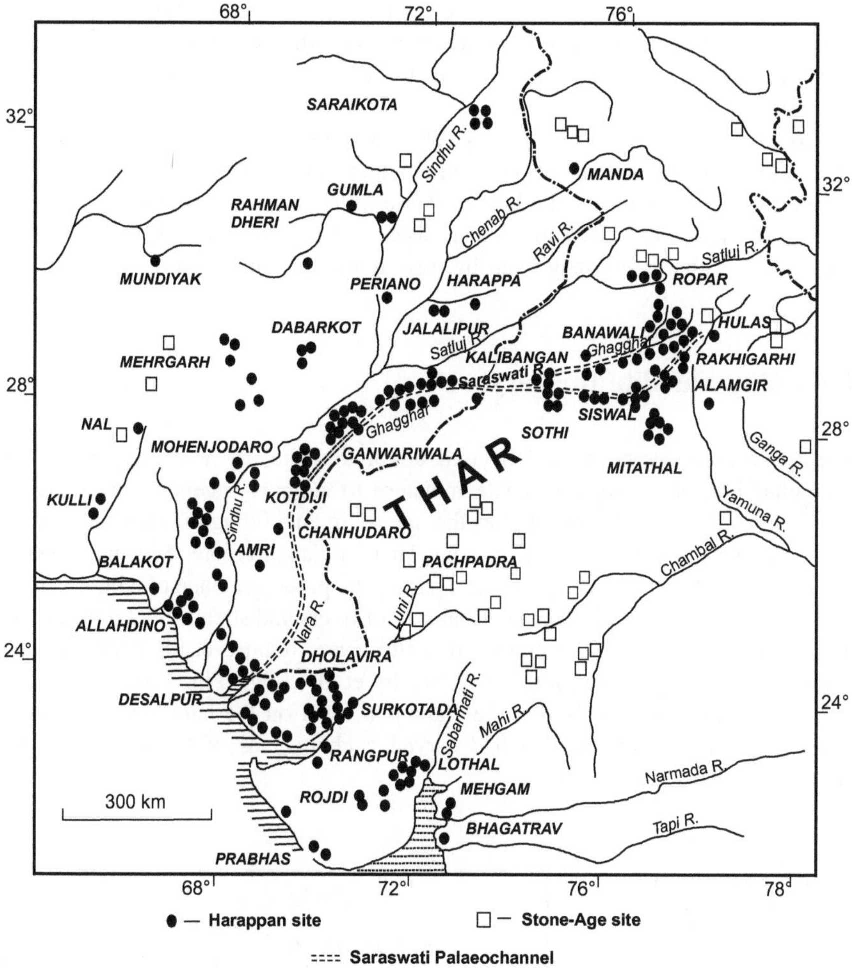

Fig. 1.2 Arid swathe of land west of the Arāvali Ranges is dotted with settlements of the Stone Ages and of the Harappan Civilization, the latter spanning the time from around 5500 to 3300 yr B.P. Note that the settlements are seen upto the sea coast in Gujarat (After Valdiya 2002, based on various sources)

Populated Land It Was

There is yet another amazing thing in this dry land. There are quite a few linear arrays of clusters of prehistoric settlements along the banks of dry but anomalously wide channels that carry, but rarely, only the flood waters during the monsoon season. The archaeologists have identified these settlements, as already stated, with the Harappa Civilization that spanned the time from around 5500 to 3300 yr B.P. (Fig. 1.2). People could not have lived long in a land that did not have water.

The existence and persistence of more than 2000 settlements over thousands of years under condition of aridity implies that water did flow round the year in the now dry channels. In other words, the dry wide channels were the courses of perennial rivers.

From the rich harvest of archaeological finds and the layout of towns and designs of buildings in the settlements, it is more than apparent that the people of the Harappa Civilization during the Mature Phase were progressive and lived a buoyant life full of artistic appreciation, elegant tastes and environmental sympathies. The implication is that rivers they lived by were bountiful with good discharge.

Revealing Satellite Imagery

The Landsat imagery revealed beyond doubt that the dry wide channel of the Ghagghar indeed contains adequate moisture to support vegetation over ground (Fig. 1.3 and Plate 1.1). This implies that the sediments filling the channels contain adequate water. Landsat imagery and later Indian Remote Sensing imagery (IRS-1C) established the Ghagghar as a part of the palaeodrainage of a big river and also brought out multitude of palaeochannels of rivers and streams that once existed in this land between the Yamunā and Satluj rivers (Ghose et al. 1979; Kar and Ghose 1984; Gupta et al. 2004, 2008; Bhadra et al. 2009) (Fig. 1.3).

There is thus little doubt that the Harappan lived on the land well-watered by rivers including a big one that had carved a channel as wide as 6–8 km—the Ghagghar–Hākrā (Lal 2002).

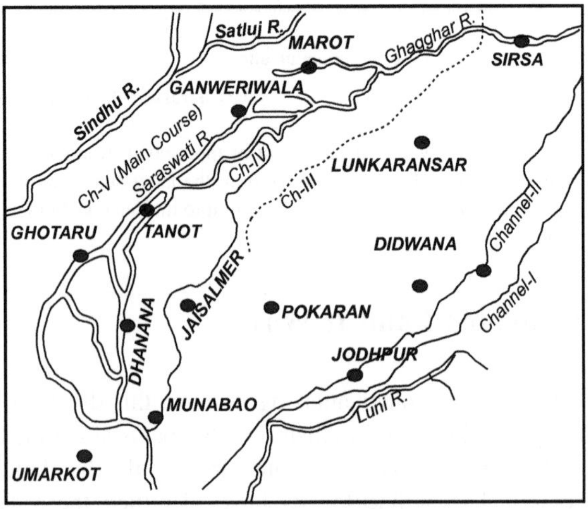

Fig. 1.3 Palaeochannels in southwestern Rājasthān as revealed by IRS imagery. Channel V represents the main course of the Saraswatī. Channel IV the alternative course. Channel III is an abandoned course of the Saraswati. The channels shown as 1 and 2 and are not related to the Saraswatī, but represent the palaochannel of the Luni River, an altogether different drainage system (After Gupta et al. 2008)

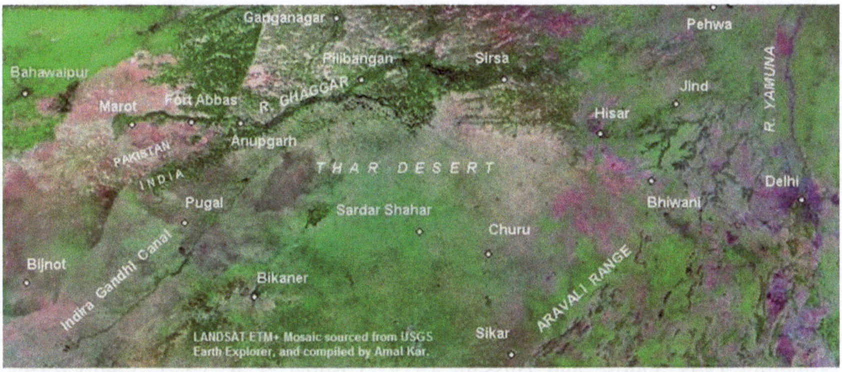

Plate 1.1 Landsat ETM + FCC mosaic brings out the Ghagghar River, which is characterized by very moist sediments, supporting greenery. Notice the bifurcation of the river in the western end–one branch of which disappears under desert sands at Marot and the other at Beriwali (*Courtesy* Dr. Amal Kar)

The Ghagghar of Haryānā is known as the Hākrā in Cholistān and as the Nārā in Sindh (Cholistān and Sindh are in eastern Pākistān). Between Anupgarh and the Rann the measured width of the channel is 4–10 km (A.K. Gupta, personal communication, 2015). The great width of channel, coupled with thickness as much as 30–50 m of the fluvial–alluvial sediments deposited by the river implies that it once carried voluminous discharge of water (Raghav 1999; Courty 1995; Saini et al. 2009). It must have been a great river indeed. That mighty river at present is reduced to the Ghagghar–Hākrā–Nārā river in which only flood water flows, but rarely.

What was that great river?

Identity of the Mighty River

In the foothills of the Siwālik (northern Haryānā) there is a petty rivulet named *Sarsuti*. Originating at Rāmpur–Herian, it flows southwest and joins a comparatively bigger stream the Mārkandā that drains the Siwālik Hills. Further downstream at Rasulā it meets the Ghagghar River, which originates in the Morni Hill in the Siwālik. About 25 km south of Patiālā at Shatrānā, the Ghagghar is joined by another wide dry Patiāli channel that is filled with thick deposits of riverine sediments, implying that it must have once been a big river. This channel originates near Ropar, the point where the south-flowing Satluj abruptly swerves westwards. The channel of the Ghagghar beyond this confluence is extraordinarily wide—6–8 km

(Yashpal et al. 1980; Gupta et al. 2004, 2008), and remains wide upto the point it breaks into two channels west of Anūpgarh in the Thār desert (Fig. 1.3).

Significantly, at Sirsā on the bank of the Ghagghar stands a fortress called "Sarsutī". Now in derelict condition, this fortress of antiquity celebrates and honours the river *Sarsutī*, the main tributary of the Ghagghar. It is therefore plausible to conclude that once upon a time the Ghagghar was known as "Sarsutī". The word "Sarsutī" is a corruption of "Saraswatī".

Was the river Ghagghar, on the banks of which hundreds of Harappan settlements were located, called Saraswatī in the past?

A growing number of scientists and archaeologists believe that the Ghagghar does represent the Saraswatī River of the yore (Oldham 1893; Raikes 1964; Wilhelmy 1969; Ghose and Hussain 1979; Yashpal et al. 1980; Valdiya 1996, 2002, 2013; Rao 1991; Mughal 1995; Courty 1995; Bisht 1998, 2013; Possehl 1999; Sahai 1999; Lal 2002; Gupta et al. 2004, 2008, 2011; Saini et al. 2009; Sinha et al. 2012; Clift et al. 2012; Giosan et al. 2012; Mitra and Bhadu 2012).

Long ago, the composers of the *Rigved* described a river abounding in water *maho arnāh Saraswatī* (*Rigved*, 1.3.12) originating in the Himālaya and emptying itself in the ocean (*shuchiryatī giribhyah ā sumudrāt*) (*Rigved* 7.95.2). It was the largest of the seven rivers of the country then called *Saptsindhav* (*Rigved* 1.35.8; 8.85.1) the land of seven rivers (Fig. 1.4).

It was this land *Saptsindhav* where the seven rivers nurtured the Harappan Civilization from the time quite well before 5500 yr B.P. to sometime after 3300 yr B.P.

This river Saraswatī is described lucidly not only in the *Rigved* but also in the *Purāns* and in the epic *Mahābhārat*, the latter two authored by Krishna Dvaipāyan 'Vyās'—a legendary writer who perhaps inspired a chain of successive writers who held the title of 'Vyās'.

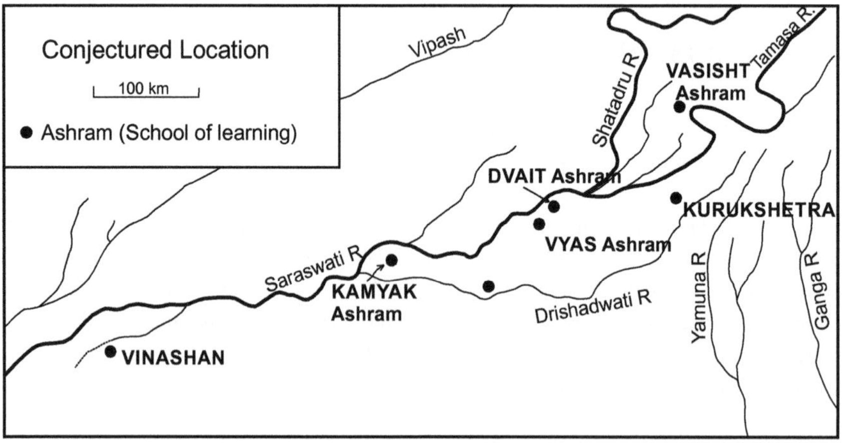

Fig. 1.4 Location of hermitages (*āshrams*) and schools of learning as described in the Purāns and the epic *Mahābhārat* (After Valdiya 2012)

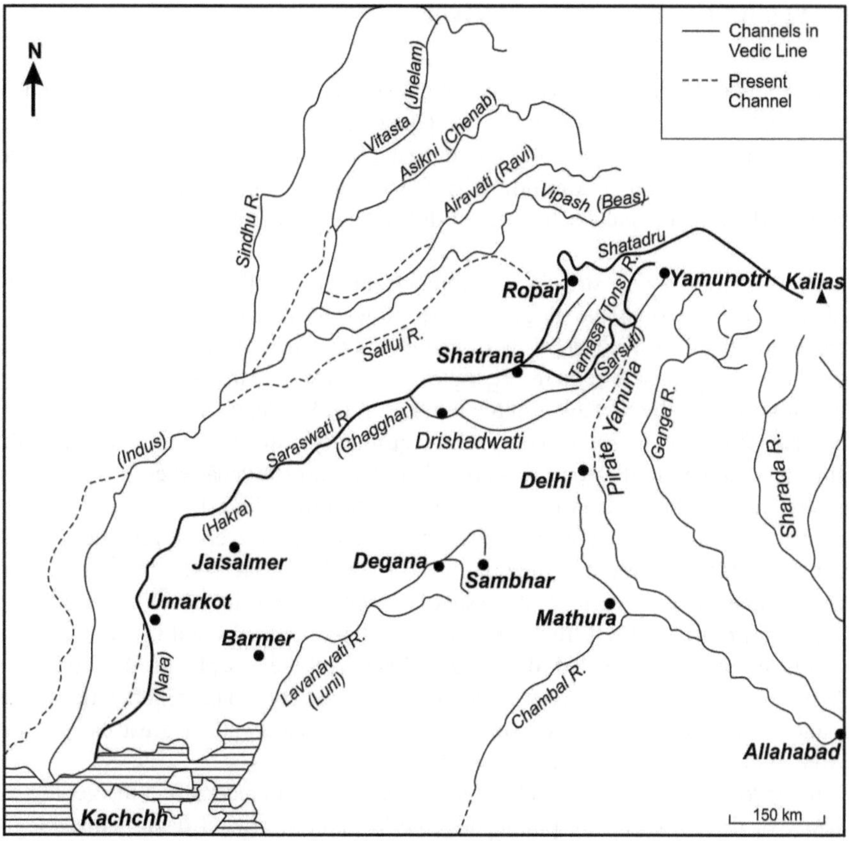

Fig. 1.5 Legendary Saraswatī, formed by joining together of the Tamasā (Tons) and the Shatādru (Satluj) rivers, flowed through the land between the Sindhu and the Gangā floodplains. It is now represented by the dry streams Ghagghar known as Hākrā in Cholistān (Pākistān) and Nārā in Sindh (Pākistān) (After Valdiya 1998)

After watering the terrain of Kurukshetra (Fig. 1.4), the Saraswatī turned westwards and flowed through such forests as the holy and celebrated Kāmyakvan, Aditivan and Dvaitvan (*Vāman Purān*, 33, verses 1–4). This is indeed the situation at present (Figs. 1.4 and 1.5). The Sarsutī turns westward near Pehowā near Kurukshetra, meets the Ghagghar, then flows past Sirsā and Anūpgarh to Mārot. Such great Harappan towns as Bhirrāna, Banāwali, Kālibangan and Ganweriwālā are located on the palaeochannel of the Saraswatī, that is the Ghagghar bank. That this river Saraswati discharged into the ocean is clear from Balarām's travelogue given in the *Mahābhārat* (Shalya Parv, 35) (*Samudram paschimam gatvā Saraswatyabdhisangam*).

It Was a Himālayan-Born River

Quite a few geologists, archaeologists and historians believe that the Saraswatī, like its present-day relic now known as the Ghagghar, was formed from the contributions of multitude of streams originating in the southern flank of the Siwālik and its foothill belt, and that it was primarily fed by heavy monsoonal rains. In northwestern Haryānā the palaeochannels of the Ghagghar are known as Sukru, Sotra or Rangoi.

However, the many streams springing from the Siwālik could not have contributed on their own the large volume of water that the channel carried and emplaced 30–50 m thick deposits of sediments throughout the extent of channels and their floodplains. It is therefore logical to surmise that the source of the mighty river was beyond the Siwālik—in the interior of the Himālaya. It was suggested that the Tamasā (old name of the Tons) with its tributary now known as Yamunā in the Himālaya, flowed in the southwesterly direction onto Haryānā where it was joined at Shatrānā by the Shatadru (now known as Satluj), which then flowed straight south from Ropar. The combined discharge of these two rivers at present is of the order of 2900 million cubic metres. In the Early Holocene when the rainfall was two to three times more than at present (Bryson and Swain 1981), the combined discharge must have been truly voluminous. Along the banks and in the floodplain of this mighty river settled down and flourished the people of the Harappan Civilization, as testified by over 2000 archaeological sites. The beneficiaries of the bounty of the river quite understandably held the Saraswatī in great esteem and wrote hyms and panegyrics on its splendor and greatness.

In the following chapters, endeavour has been made to reconstruct the geological history of the Saraswatī by putting together a mass of critical data, painstakingly gathered over several decades by archaeologists, geologists, geomorphologists, geohydrologists, geochronologists and remote-sensing specialists.

Chapter 2
Geomorphological Layout of the Saraswatī Land

Land of Triple Provinces

The drainage of the two rivers that once constituted the two branches of the legendary Saraswatī River of prehistoric antiquity encompassed three radically different physiographic–geological provinces. One of the branches, the Satluj (once called Shutudri or Shatadru) arose in southern Tibet beyond the Himālaya, the other the Tons (earlier called Tamasa) emerges in the ever-snowy Himādri or Great Himālayan domain, and the third the Chautang (Drishadwati of the Rigvedic times) drained the northern foothill belt of the western Indo-Gangetic Plains (Plate 2.1, Fig. 2.1). In its middle reaches the river traversed a large swathe of alluvial plain, and then passed through the dreary landscape of sand dunes of the Thār Desert, before ending in the Rann of Kachchh, once a part of the realm of the Arabian Sea.

Mountain Province

Standing high like a colossal sentinel, the Himālaya province is divided into four terranes or subprovinces; one altogether different from the other in respect of landforms, lithological makeup, structural design and tectonic history (Fig. 3.2). These are the *Tethys Himālaya* in the north, the *Great Himālaya* or Himādri, the *Lesser Himālaya* and the *Siwālik* in the south. North of the Himālayan province, the Tibetan landmass is an undulating terrain of peneplaned plateau 3600–7500 m above the sea level (Plate 2.4, Fig. 3.2). In the Tibetan belt the eastward-running Brahmaputra and the northwestward-flowing Sindhu have carved out their valleys and built floodplains of gentle topography. These two rivers emerge from a terrain of domal upwarp capped by the many-splendoured Mount Kailās (6714 m). South

K.S. Valdiya, *Prehistoric River Saraswati, Western India*,
Society of Earth Scientists Series, DOI 10.1007/978-3-319-44224-2_2

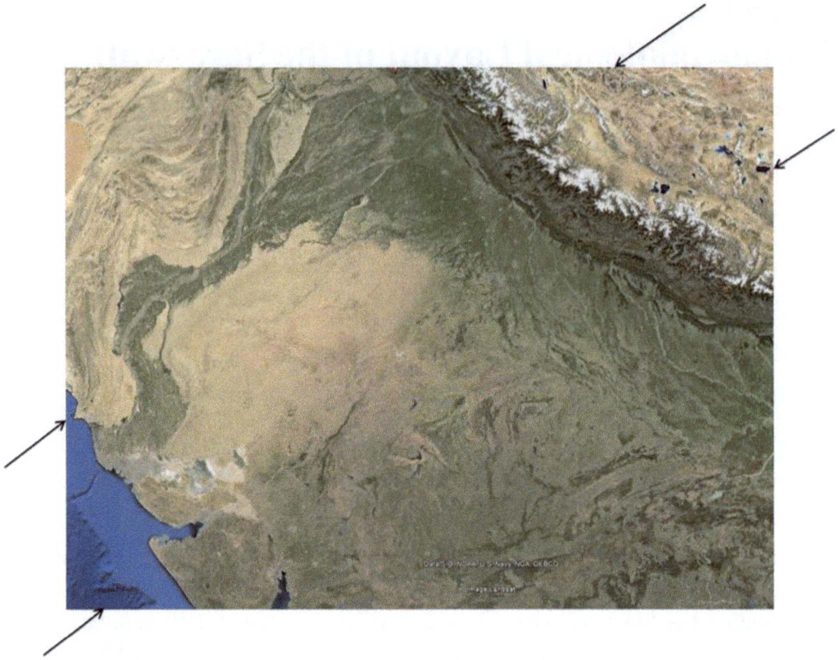

Plate 2.1 Satellite imagery shows the three geologically different and physiographically contrasted provinces—the trans-Himālayan Tibet, the geodynamically mobile Himālaya, the Indo-Gangetic Plains and the Arāvali domain of the Indian Shield—through which once the legendary Saraswatī River of antiquity flowed. The *arrows* mark the domain of the Saraswatī basin (*Source* Google Earth)

of this topographic eminence are the twin lakes—Mānsarovar and Rākshastāl (Plates 2.1 and 2.2). It is a spot west of the Rākshastāl that is the source of the Satluj (Shatadru and still earlier Shutudri).

The Tethys Himālaya is a rugged terrain remarkable for its fantastically sculptured landforms in sedimentary rocks. Bereft of vegetation on the whole, this desolate domain is a cold desert populated extremely sparsely, that too only in isolated places in valleys where clumps of trees have found foothold. Glaciers abound, and rivers have cut deep V-shaped valleys across the ranges of sharply rising peaks. To the south of the Tethys subprovince stretches in a great sprawl the lofty Great Himālaya—also called Himādri. Perenially capped by snow and ice, the rugged ranges of the Himādri rise 3000 to more than 7500 m high. Bandarpunchh (6320 m) and Leo Pargial (6770 m) in the west are amongst the tens of high peaks of this subprovince. From one of the glaciers of this domain (the Har-ki-Dun) emerges the Tons (Plate 2.3) and from another its tributary the Yamunā. The Tons was earlier known as Tamasā. And through Shipkilā gorge and past the Leo Pargial flows the Satluj. The Himādri has extremely youthful and forbiddingly rugged

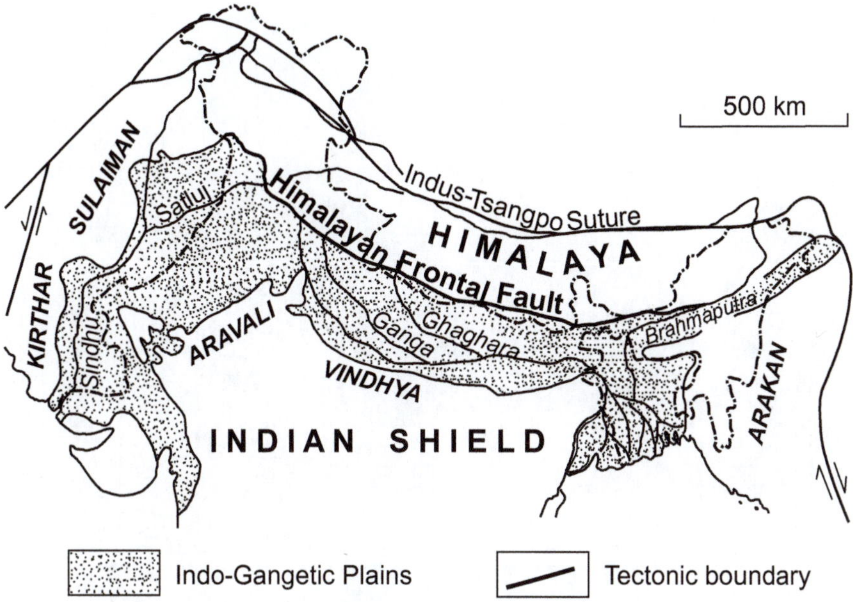

Fig. 2.1 Sketch map shows the floodplains of the Saraswatī and Sindhu rivers in the western IndoGangetic Plains

Plate 2.2 The Rākshastāl, south of Mount Kailās. Close to this lake is the source of the western branch of the Saraswatī, known as Shatadru or Satluj (Photographed by Anup Sah, Nainital)

Plate 2.3 Har-ki-Dun nestling on the southern flank of the Great Himālaya is the source of the eastern branch of the Saraswatī—the Tamasā, today known as Tons (Photographed by Shree Niraj Pant). The lower picture shows a tributary of the Tons river in very upper reaches

topography of the lithotectonic block that is made up of high-grade metamorphic rocks and gneissic granites. The southern face of the Great Himālaya is broken by high precipitous scarps. Through very deep gorges across the high ranges flow the rivers Satluj, Tons and Yamunā, churning violently through the narrow canyons.

The Great Himālaya overlooks the subprovince of the Lesser Himālaya, the elevation of which ranges from 600 to more than 2500 m (Fig. 3.2). The rocks that make up the mountains are sedimentary rocks, thrust over low-grade metamorphic and volcanic rocks. In contrast to the other subprovinces, the Lesser Himālaya is a relatively mild terrain with gently sloping mountain sides and rounded summits. Streams and rivers flow in their comparatively wide, winding valleys, which become suddenly narrow gorges on crossing faults that are very common. In such belts of crossing, the landscape is very rugged. The mountains wear carpets of rich soil that once supported dense forests but are now bereft of the sylvan cover over greater part of the land. This subprovince happens to support comparatively dense population. In the south is the Siwālik subprovince that is made up of exclusively sedimentary rocks. The Siwalik ranges rise 250 to 1600 m, forming the southern front of the Himālaya. The rugged ranges are commonly broken by south-facing scarps, while on their northern steeper slopes rush down streams through unending cascades. Then there are long flat stretches of thick gravel deposits within the synclinal valleys of the rugged Siwālik hills. These are called *dūns*, such as the

Plate 2.4 Satellite imagery shows the mountain domain (Himālaya and Tibetan plateau) of the Saraswatī. *Arrows* indicate the areas of the sources of Shatadru (Satluj) and the Tamasā (Tons) (*Source* Google Earth)

Pinjaur Dūn and the *Pāontā Dūn* through which the Yamunā–Tons flow. The Siwālik is covered with dense forests and is sparsely populated.

Foothill Belt

At the foothills of, the Siwālik the 360–240 m above sea level, the northernmost belt of the Indo-Gangetic Plains embodies a fringe of coalescing fans of gravelly deposit over a width of 20–40 km, some of which extend 7–15 km southwards from the immediate foothills. This happens to be the youngest stratigraphic unit but older than some terrace alluvium of the province of the Indo-Gangetic Plain. This unit is on the average 7–10 m thick; and known as *Bhābhar*. The gravel deposits were formed by rivers and streams coming from the mountains and dumping their heavy loads as the gradient of the ground decreased. These have been dated >80,000 ± 500 yr B.P. to <51,000 ± 300 yr B.P. (Shukla and Mujtaba 2015). Streams have cut rather wide and sharp channels characterized by 3700 ± 100 yr old terraces (Plate 3.1), more than 3 m above the stream beds.

Digital elevation model constructed from SRTM and ASTER satellite data, combined with fieldwork between Yamunā and Satluj rivers brought out a large, wide fan of gravelly deposits—earlier named as Terminal Fan (Mukherji 1976)—the surface of which dips southwards (gradient 20–30 cm/km) *and characterized by diverging abandoned palaeochannels of a river* (Srivastava et al. 2006). This unit shows highest drainage density anywhere in the Saraswati land. Within the tract of the frontal Siwālik range there are spectacular 30–100 m high vertical river-cut cliffs and the seasonal streams characterized by terraces (Chaudhuri 2008a). And in the immediate foothill belt there are several short and narrow incised seasonal water courses, the main one being the Sarsutī, the Mārkanda, the Dāngri, the Ghagghar and the Patiāli (Fig. 2.2).

A very remarkable feature of this belt is that the water-divide between the Gangā and the Sindhu does not coincide with the highest point of the interfluves between the Yamunā and Satluj rivers—it is 15–93 km away to the west (Srivastava et al. 2013). It seems that the ground made up of southwestward-expanding alluvial fan and characterized by diverging drainage system is arched nearly 3 m up. The surface profiles (Srivastava et al. 2013) across this huge fan (Fig. 2.3) indicate that most of the seasonal streams have carved extraordinarily wide water courses, within which the active streams occupy narrow incised channels (Plate 3.1).

West of Ghagghar (Fig. 2.2), five wide dry channels, known as Patiāli, Wah, eastern Naiwāl, central Naiwāl and western Naiwāl, emerge from the point at Ropar —the point where the southward flowing Satluj River abruptly swerves west. And west of the Satluj in the area of the Beās–Rāvi interfluves, 0.5–5 m thick layers of aeolian sands are intercalated with more than 150 m thick multicyclic sediments of the unit known as *Bhangar* in Indo-Gangetic Plains (Rubi and Punj 1997). The presence of aeolian sands in northern part of the western Indo-Gangetic Plains

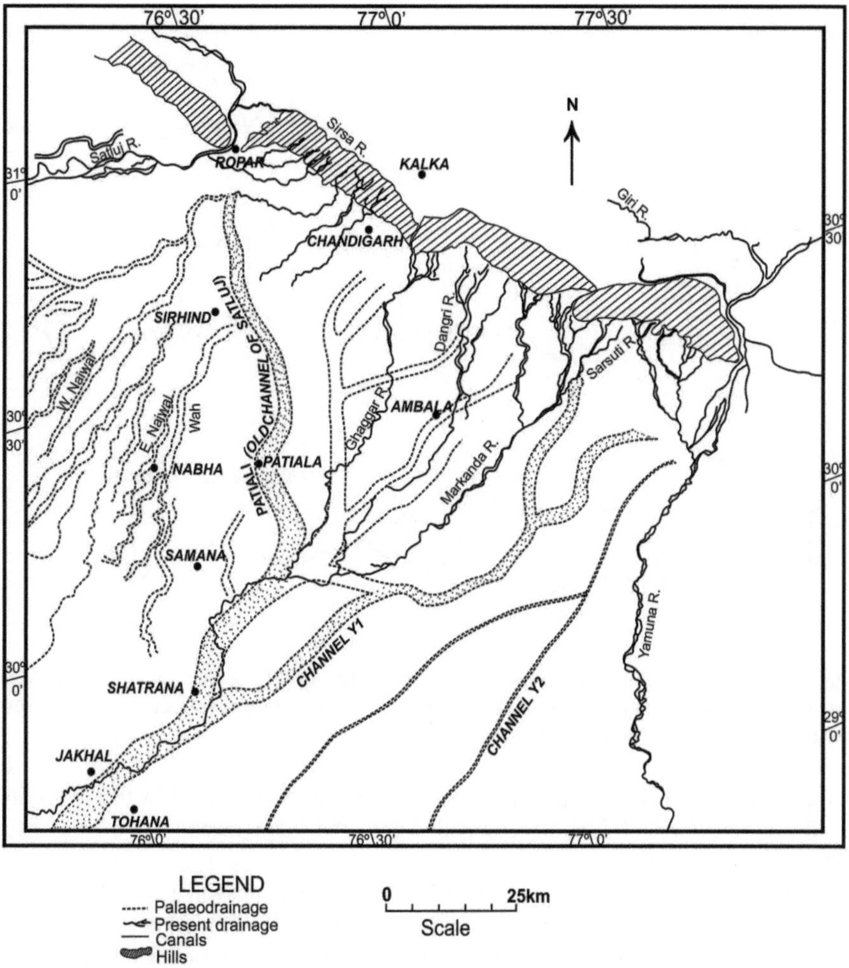

Fig. 2.2 Siwālik-born rivers and streams south of the Siwālik ranges. Most of these are ephemeral streams and associated with palaeochannels (After Yashpal et al. 1980)

implies the great sway of the dust storms that rise from the Thār Desert situated far to the southwest.

Floodplain in Middle Reaches

The floodplain (240–120 m above the sea level), through which the Ghagghar and its tributaries flow, gently slopes first southwards then westwards and finally southwestwards. In the Ambālā–Ludhiānā tract, the elevation is 240 m and then

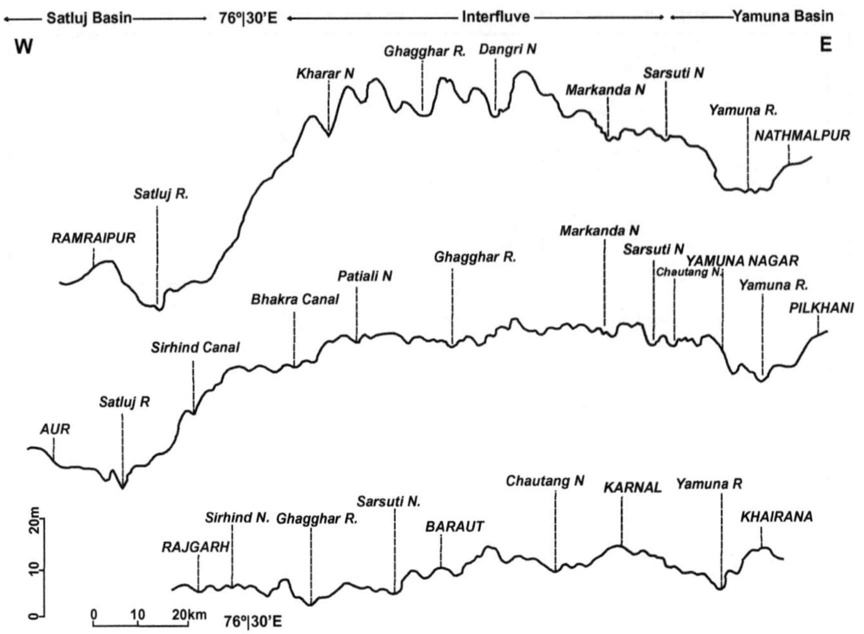

Fig. 2.3 Profiles across the large alluvial fan south of the Siwālik foothills show the wide water courses of seasonal streams and the narrow channels of active streams (vertical scale very much exaggerated) (After Srivastava et al. 2013)

progressively decreases to just a few metres above the sea level in the low-lying stretch of the Great Rann of Kachchh where the Nārā reach of the river ends (Plate 2.1, Fig. 2.1).

East of the Ghagghar basin is the NE–SW trending Arāvali front, the western foot belt of which is covered by discontinuous mantle of sediments derived from the hill ranges. In the Sirsā area there is a depressed tract—seven metres lower than the general elevation of the floodplain. The bluffs at Razābād and Bajika is indeed steep cliff that drops 7–8 m within a span of just 30 m. Not only the depressed ground, the Ghagghar channel in the Hisār–Sirsā tract is characterized by two paired fluvial terraces, respectively 10 and 3.5 m above the present river-bed (Singh et al. 1988). The higher terrace is more conspicuous. The terraces imply uplift of the ground through which the river flowed.

Downstream of Sirsā the river channel is 4–6 km wide. The northern limit of the floodplain is defined by a sharp palaeolake and the southern limit is buried under the aeolian deposit. Between Tohana and Ellenābād, there is a 2–10 km wide palaeochannel to the south of the Ghagghar (Saini et al. 2009).

Lower Reaches: Desert Terrain

West of the flat alluvial expanse, the land is covered by wind-blown sand heaped up as ridges and dunes. Varying in height from less than a metre to as much as 10 m, and in length a few metres to more than 5 km, the dunes are mostly of linear and parabolic shapes (Kar 1993, 2014a) forming long ridges as well as crescent-shaped mounds (Plate 2.5; Figs. 2.4 and 2.5). In addition to the dunes there are sand sheets of more than 5 m in thickness. The dunes and the sand sheets form the undulating landscape of the Thār Desert.

The sands of the desert represent sediments initially deposited by rivers, later churned and blown off by storms and swept away and heaped elsewhere (Fig. 2.5a). According to Dhir and Singhvi (2012) much of the Quaternary period in the Thār being arid, the oldest aeolian sand date nearly 200,000 yr B.P. Before the Last Glacial Maximum, the major emplacement of desert sands took place in four major

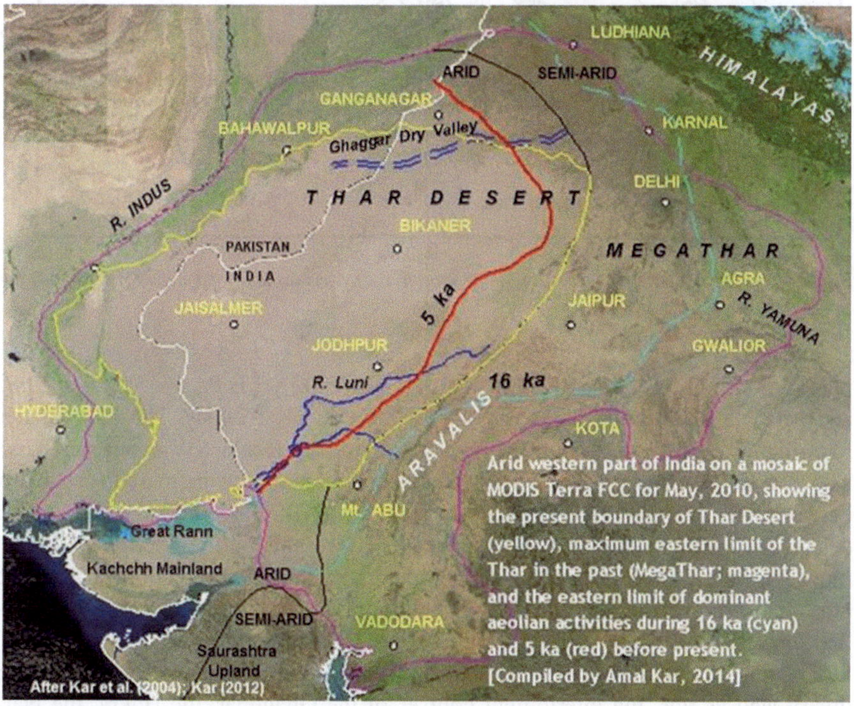

Plate 2.5 MODIS-Terra FCC of western part of the Indian subcontinent draped with the present Thār Desert (*yellow boundary*). The maximum eastern limit of the Thār in the part is shown by magenta line. The eastern limit of the dominant aeolian activities 16,000 years ago (*red line*) and of semi-arid region 5000 years ago (*black line*) (After Kar et al. 2004; Kar 2012)

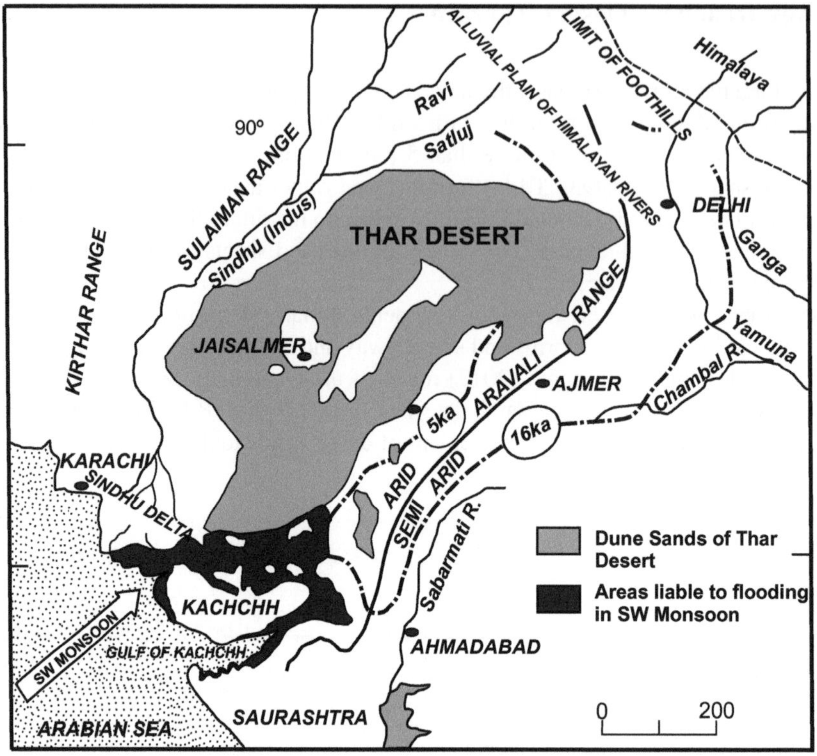

Fig. 2.4 Extent of the Thār Desert of sand-covered rocky terrain and sand dunes (From Dhir and Singhvi 2012)

phases—during 115,000–100,000 yr B.P., ~75,000 yr B.P., ~55,000 yr B.P., and between 33,000 and 2500 yr B.P. (Singhvi and Kar 2004). The Last Glacial Maximum (24,500–18,000 yr B.P) was a period of feeble SW monsoon wind. The SW monsoon wind peaked up after the dune-building increased, especially between the intervals 16,000–10,000 yr B.P. and 14.000–10,000 yr B.P., Thār (Plate 2.5, Fig. 2.4) then expanded beyond its present limit. A period of higher monsoon rainfall followed, reaching the maximum during 7000–6000 yr B.P; and after that aeolian activity was confined within the desert; particularly during 5000–3500 yr B.P. and 2000–8000 yr B.P. (Kar et al. 2004).

The huge mass of aeolian sands extends almost up to nearly the coast. In such a dreary land as this, the intermittent courses of the Ghagghar, the Hākrā, the Nārā become increasingly unclear. However, one can see the trace of the river where the dunes are less numerous (Flam 1993). The Hākrā course from Fort Abbās to Fort Derāwal is marked by a depression partially filled with sediments. Southeast of Fort Derāwal the Hākrā course is lost. A river that originated in the Great Himālaya (Himādri) and flowed with full force through the Ghagghar course all the way up to

(a)

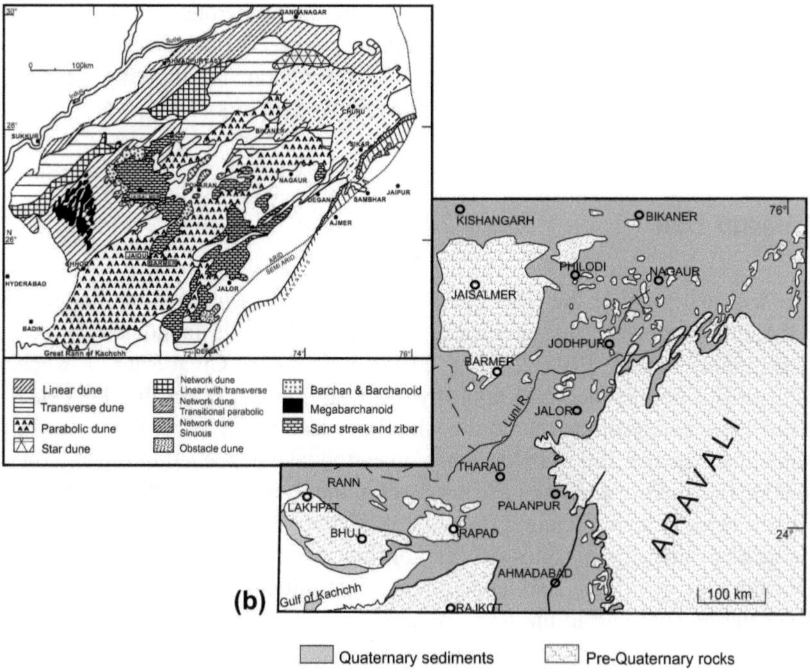

(b)

Fig. 2.5 **a** Types of sand dunes in the Thār Desert (After Kar 1996). **b** Lower reaches of the Saraswatā encompass inliers amidst sand of wind-worn Pre-Quaternary rock. *Broken lines* shows international boundary (After Bakliwal and Wadhawan 2003)

Cholistān (Pākistān) would not have abruptly ended (dried up) without flowing further downstream. There is no denying that the downstream reach is largely hidden under the pile of desert sands. However, the graphic descriptions given by explorers, engineers, geographers and archaeologists (Makeson 1844; Raverty 1892; Stein 1942; Mughal 1995; Panwar 1986) who worked extensively in that swathe of land (Cholistān and Sindh) provide credible account of the channel of the great river.

Analysis of aerial photographs and satellite imagery reveals existence of much older dysfunctional and disorganized drainage system in the Bikāner–Jaisalmer region (Kar and Ghose 1984; Kar 1988; Ramasamy et al. 1991; Sahai 1997; Rajawat et al. 1999). By dysfunctional drainage it is meant that there are palaeochannels without flowing water and they form a network of complex drainage.

In western Rājasthān, there are inliers of worn out older rocks (including Precambrian) jutting out above the sandy waste of the *Marusthali* (desert) (Fig. 2.5b). East of the Thar Desert—in the rocky terrain between the foot of the

Arāvali and the Bikāner–Jaisalmer belt—there are numerous salt lakes and playas. The playas and salt-encrusted clay deposits represent the lakes that have dried up. These playas and salt lakes occur in a riverine domain, implying their origin to blockages in the courses of the once flowing river (Ghose 1964; Kar 1988, 1990).

Coastal Belt

The terminal point of the Ghagghar–Hākrā–Nārā (Saraswatī) is the flat expanse of the salt-impregnated, salt-encrusted, marshy land that is inundated by sea water during rainy season (Plate 2.6; Fig. 2.6). Known as the Great Rann of Kachchh, it is a sunken tract encroached upon by sea water, as evidenced by thick tidal sediments. It is believed that the floor of the Rann sank a number of times, the last time sometimes around 2500 yr B.P. when an earthquake happened in this region (Merh 2011). The sinking and rise of water level in the Rann of Kachchh is attributed also to sea level changes (Gaur et al. 2013; Mathur 2002). It seems that there was interesting interplay of land–sea relationship as transgression followed regression (Hashimi et al. 1995) and concomitant intermittent uplift of the landmass of Kachchh and sinking in the Rann region.

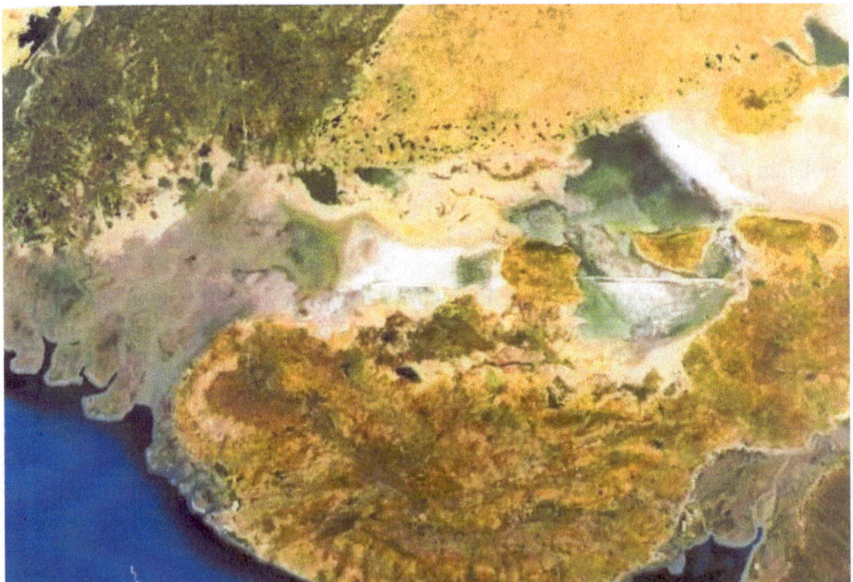

Plate 2.6 Satellite picture of the region where the Saraswati ends into the part of the sea today represented by the Great Rann of Kachchh. The Rann is flanked in the south by the island of Kachchh (*Source* Google Earth)

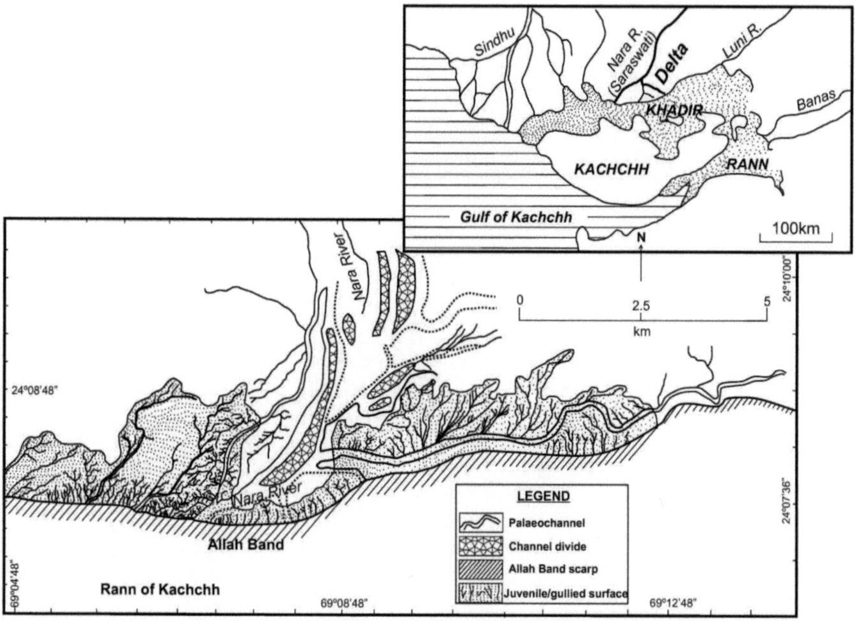

Fig. 2.6 Delta of the Nārā River—the terminal reach of the ancient Saraswatī—against the Allāhband high in the Great Rann of Kachchh (After Tyagi et al. 2012). *Inset* Deltas of three rivers that emptied into the Rann Sagār—the Rann that was once under sea water (from Ngangom et al. 2012)

The northern part of the Rann comprises three large deltas—the western one attributed to the Sindhu that then flowed into the depression (Flam 1993), the middle delta built by the Nārā (i.e., Hākrā, i.e., Ghagghar), and the eastern one by the Luni River (Fig. 2.6 inset and Plate 2.6), the Luni draining the southwestern flank of the Arāvali (Sharma and Bhadra 2012). It may be mentioned that quite many workers have concluded that the Rann of Kachchh is an extension of the Gulf of Kachchh—or an inlet of the Arabian Sea (Siveright 1907; Flam 1993).

Chapter 3
Structural Framework and Tectonics of the Saraswatī Basin

Tectonically Resurgent Himālayan Reach

The Tamasā (Tons) branch of the Saraswatī, springing from the Har-ki-dūn glacier in the Great Himālaya subprovince (Plate 2.4), crosses three major thrust zones—and three litho-tectonically distinctive and phyisographically contrasted terranes—before descending onto the plains in southwestern Uttarākhand (Fig. 3.1). Pronounced in their structural characteristics such as severe deformation, shearing and mylonitization (milling) of rocks, these three thrust zones of regional dimension demarcate the southern boundary of the three subprovinces—the Great Himālaya (*Antargiri*), the Lesser Himālaya (*Bahargiri*) and the Siwālik (*Upagiri*).

The *Main Central Thrust* defines lower boundary of the Great Himālaya (Himādri) terrane made up of high-grade metamorphic rocks such as garnet-biotite-muscovite schist, garnetiferous quartzite and kyanite–sillimanite gneisses with subordinate calcsilicate gneisses of Precambrian antiquity (Figs. 3.1 and 3.2). The metamorphic assemblages commonly in the upper part are intruded by 21–22 million-year-old (early Miocene) granites of anatectic origin. The Great Himālaya is a huge monolith with undeciferable mega structures, but split up into a number of lithotectonic slabs by thrust planes.

The Lesser Himālaya is a subprovince (Figs. 3.1 and 3.2) of marine sedimentary rocks of Precambrian (1700–525 m.y.) ages. The preponderant rocks are quartzites, lithic quartzites and slates intimately associated with basic volcanics (basalt) and intrusive sills (dolerite) succeeded by carbonates (dolomite, dolomitic limestone and limestone) which are associated with carbonaceous slates particularly in the upper part. There are lens-shaped bodies of intraformational conglomerates in the lower and upper horizons, but they are of limited dimension. These sedimentary rocks have been hardened by load metamorphism. The rocks are thrown into anticlinal and synclinal folds, quite many of them overturned or even recumbent. The succession of crystalline rocks was thrust over the sedimentary pile. The last

K.S. Valdiya, *Prehistoric River Saraswati, Western India*,
Society of Earth Scientists Series, DOI 10.1007/978-3-319-44224-2_3

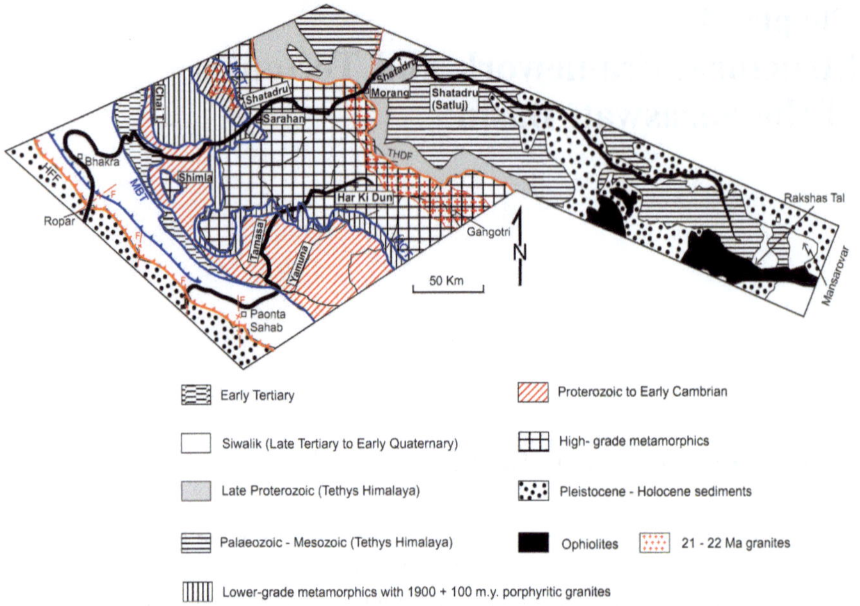

Fig. 3.1 Geological sketch map of the Saraswatī Basin in the Himālaya showing broad lithostratigraphic divisions of rock formations and the terrane-defining prominent thrusts of regional dimension (After Thakur 1993)

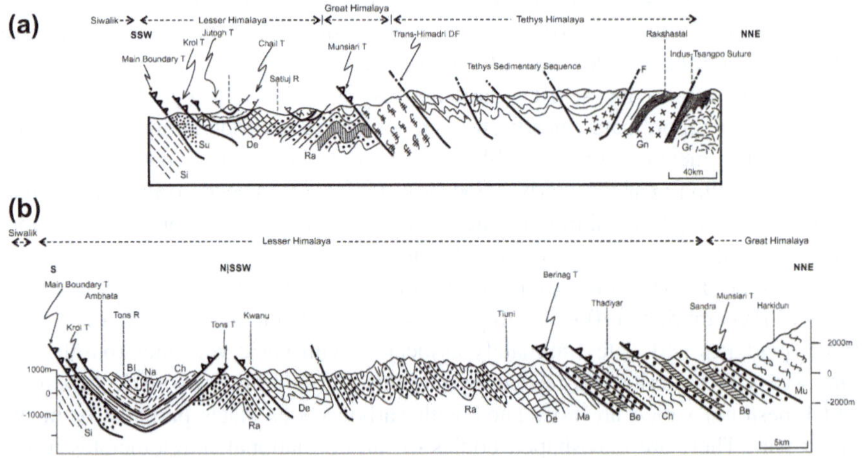

Fig. 3.2 Cross sections of the mountains across which the Satluj and the Tons flow. The sections show terrane-defining thrusts of regional dimension and the broad divisions of the lithostructural terranes **a** After Vadiya (1980) **b** After Auden (1935)

phase of folding took place after the sedimentary succession was thrust over and thus overlain by piles of lower-grade metamorphic rocks which were intimately associated at the base with highly mylonitized 1900 ± 100 million-year-old porphyritic granites and intruded by 525 ± 25 Ma granites-granodiorite bodies. These crystalline rocks were brought south from their root somewhere in the north, and now occur as thrust sheets (nappes) forming synclinal mountain ranges (Fig. 3.2).

In the south, the *Main Boundary Thrust* (Fig. 3.2) delimits the boundary of the Lesser Himālaya against the Siwālik. Made up of Later Tertiary and Early Quaternary fluvial–alluvial sedimentary rocks, the Siwālik is a much folded and much faulted terrane, particularly closer to the Main Boundary Thrust. The rocks that make up the Siwālik are sandstones, shales, mudstones and conglomerates—the conglomerates occurring in the upper part of the succession. Having escaped load metamorphism, these rocks are softer and easily amenable to erosion.

The Siwālik subprovince is faulted against the flat expanse of the Indo-Gangetic Plains made up of Late Pleistocene to Holocene sediments deposited by rivers and streams in their channels and flood plains. It is the *Himālayan Frontal Thrust/Fault* that dismembers the Siwālik from the Indo-Gangetic Plains (Fig 3.3).

The Satluj (Shatadru) flows through much larger expanse of the mountain domain. Originating near the twin lakes Rākshastāl–Mānsarovar (South of Mount Kailās) (Plates 2.2 and 2.4), it flows through nearly 200-km wide swathe of the land belonging to the *Tethys Himālaya* subprovince, the Great Himālaya domain, the Lesser Himālaya, and the Siwālik, and then enters the Panjab Plain—the western part of the Indo-Gangetic Plains—at Ropar (Figs. 3.1, 3.2 and 3.3). The Tethys terrane is made up of marine sedimentary rocks ranging in age from Proterozoic to upper Jurassic. In Tibet wide stretches of the Tethys Himālayan subprovince in the Satluj catchment is covered by thick piles of alluvial and fluvial sediments of Quaternary age (Heim and Gansser 1939). The southern part of the Lesser Himālaya in the Satluj basin comprises one more lithotectonic unit—the Early Tertiary unit made up of predominatly fluvial sedimentary rocks that are thrust southwards upon the Siwālik in the south and themselves overridden by Lesser Himālayan Precambrian rocks.

The Satluj, thus, traverses two more terranes or subprovinces and two more terrane-defining thrust planes, namely the *Trans-Himādri Detachment Fault* separating and dislocating the Tethyan sedimentary pile from its basement the Great Himālaya complex of crystalline rocks, and the *Krol Thrust* that has brought the Lesser Himālayan rocks on the Early Tertiary sedimentary succession (Figs. 3.1 and 3.2).

The most significant fact about this tectonic framework of the mountain reaches of the Saraswatī is that not only are there a multiplicity of strike- and dip-faults and transverse tear faults but also terrane-defining thrusts that happen to be active. Active faults are the ones on which movements have taken place in the late Quaternary times including the Holocene. The older rocks ride over even on younger riverine terraces and on the fans and cones made up of landslide debris or screes. Quite many streams are beheaded and deprived of their headwaters. The Pleistocene–Holocene gravel deposits are tilted or uplifted, giving rise to stepped terraces. Rivers rush down the over-steepened gradients through canyons formed across uplifted fault blocks (Valdiya 2001). These and many other related developments point to geolocally recent strong movements that took place on not only the terrane-demarcating boundary thrusts, but also on many other faults and shear zones that occur within the terranes (subprovinces). The recurrent movements must have generated earthquakes with resultant collapse of mountain sides and landslides or debris avalanches. The 50-km wide belt of high seismicity just south of the Main Central Thrust is relentlessly ravaged by landslides, debris avalanches and rockfalls of spectacular proportions. These facts bear testimony to the Main Central Thrust zone being tectonically very active. Movements on the Trans-Himādri Detachment Fault is manifest in the blockage of the rivers Kāli, Eastern Dhauli, Gori and Western Dhauli in Uttarākhand. The blockage gave rise to huge lakes upstream of the impediment created due to uplift of the downstream block moving up on the fault plane (Valdiya 2001; Valdiya and Pande 2008). The palaeoseismites in the lake sediments indicate that earthquakes occurred in the Tethys terrane during 40,000–20,000 yr ago (Pant et al. 2006) and about 20,000 ± 300 yr, 18,000 ± 300 yr and 13,000 ± 200 yrs ago (Juyal et al. 2004).

Tectonically Convulsed Foothills

Both the Main Boundary Thrust and the Himālayan Frontal Thrust Fault that demarcate the Siwālik terrane (Fig. 3.3) are very active in Uttarākhand (Valdiya et al. 1984; Valdiya 1992, 2001; Vadliya et al. 1992) as evident from, among others, pronounced rightward and leftward swinging of rivers that cross them, by the truncation of cone-shaped or fan-shaped landslide- or debris-flow deposits emplaced across the fault planes, and by the occurrence of thick clay deposits laid

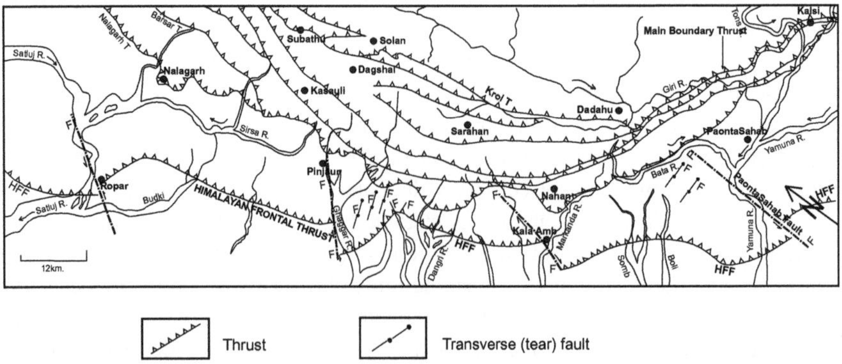

Fig. 3.3 Multiplicity of thrusts characterizes the Outer Lesser Himālaya and the Siwālik. The Siwālik subprovince is delimited by the Main Boundary Thrust and the Himālayan Frontal Thrust/Fault. It is a much-faulted terrane in the Siwālik domain (Based on ONGC Map)

down due to stream blockage upstream of the points where they cross the fault planes. A series of reverse faults with variable inclination, the *Himālayan Frontal Fault* (Figs. 3.2 and 3.3) has truncated and considerably attenuated the Siwālik subprovince. At present it is known to be much more active than the Main Boundary Thrust. At a place called Tribalan in the Yamunā valley, the dip-slip movement on the Himālayan Frontal Fault lifted up by 20–30 m the gravelly deposits of a stream containing 3663 ± 215 year-old carbonaceous matter (Wesnousky et al. 1999). A scarp bears testimony to the faulting up of the stream-bed gravel.

The outer Siwālik is equally active in Himachal Pradesh. Study of six palaeoseismic sites in the canyons of the Ghagghar, the Mārkandā (Figs. 3.3 and 3.4) and the Kosi rivers demonstrated uplift of bedrocks at the rate of 4–6 mm/yr, resulting in slip rate of 4–16 mm/yr on the fault plane inclined 20°–24° northward (Kumar et al. 2001, 2011). The A.D. 1500 earthquake that occurred in the fault related to the Himalayan Frontal Fault is attributed to 18–28 m slip (Thakur 2006). That must have been caused by a mega-earthquake or possibly many great earthquakes.

Not only that, the frontal Siwālik Range has been torn apart and left-laterally dislocated by the NNW–SSE trending Pāonta Sāhab Fault, also known as Yamunā Tear (Fig. 3.3). This wrench fault follows the valley of the Yamunā River. Flowing in a braided channel, the Yamuna is flanked by a multiplicity of paired and unpaired terraces and straths, the younger ones of the Holocene time. Significantly, the oldest terrace representing the fill of as abandoned channel is characterized by a zone of prominent soft-sediment deformation structures (Pandey and Pandey, 2015). An earthquake that occurred 800 years ago is related to this transverse fault (Joshi et al. 2005). The fault continues to be active in the present—horizontal movement of 30 cm between 1962 and 1966 (Krishnaswamy et al. 1970) and 0.7–5.7 cm between 1965 and 1976 (Rajal et al. 1986) have been recorded.

The movements on the Himālayan Frontal Fault and on the transverse, and longitudinal faults that cut the Siwālik subprovince into blocks, are manifest in the development of terraces making a step-like design formed in that period and seen lining the stream and river valleys (Plate 3.1). These terraces point to repeated uplift of the ground and concomitant acceleration of stream dissection. In the Yamunā valley for example, there are four successive terraces at the elevation of as much as 65–80 m above the river bed and dated (by OSL method) 11,000–5000 yr, 7000–4000 yr, 3000–2000 yr and less than 2000 yr respectively (Dutta et al. 2012). Needless to state, the Pāontā Dūn within the Siwālik through which the Tons (the Tamasa branch of the palaeo-Saraswatī) flows, was lifted up four times in the past 35,000 years. The situation to the east in Dehra Dūn is as much tectonically resurgent as in the Yamunā valley. A low-angle 19 km long thrust, passing through Bhuwāwālā, has warped up the Dehrā Dūn surface. This surface is dated by OSL method at 2619 ± 225 yr B.P. and by radiometric method at 2570 ± 220 yr B.P. (Jayangondaperumal et al. 2008). The development of what is called Newer Dūn due to 20–30 m uplift and 15°–20° tilting of the Older Dūn in the Dehrā Dūn

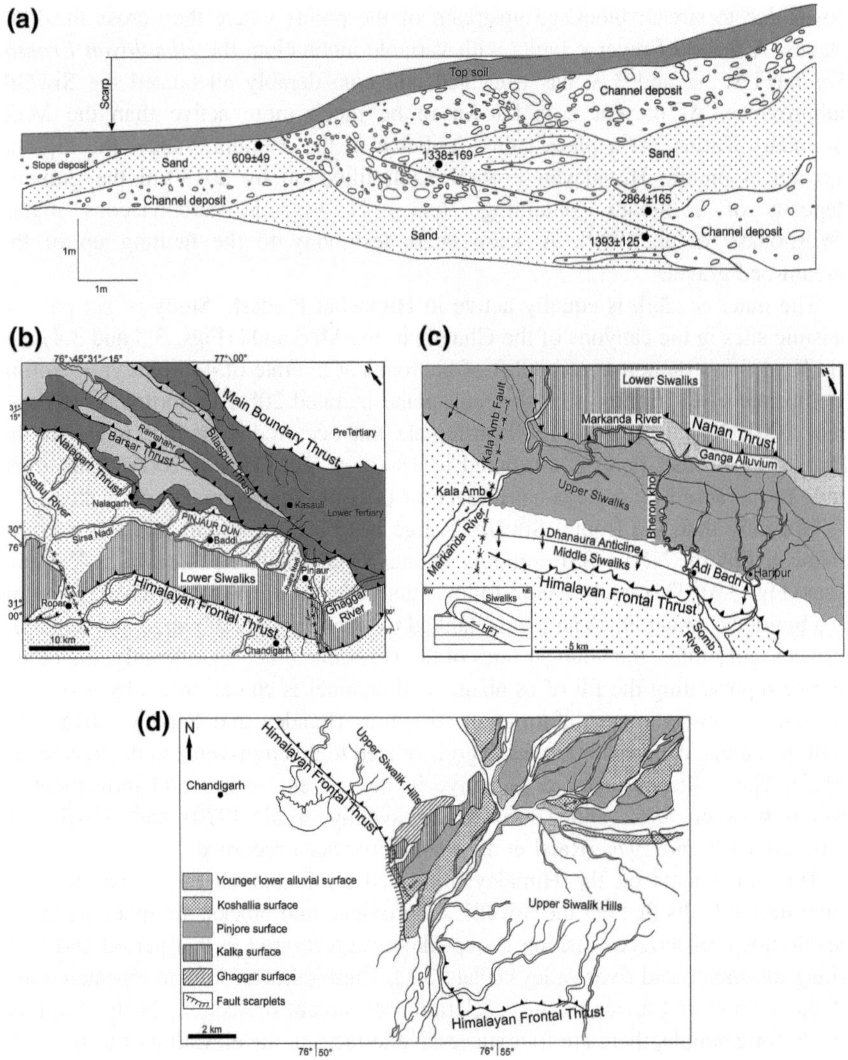

Fig. 3.4 Neotectonic activities in the foothills of the Siwālik. **a** Trench dug at Singhauli shows movement on the HFT and associated thrust planes considerably deforming the Recent sediment. **b** Tear (transverse) faults have dislocated the frontal range of the Siwālik. Rivers Ghagghar (east) and Satluj (west) have opened up their channels through the fault zones. **c** Himālayan Frontal Thrust Fault offset dextrally at Kālā Āmb. **d** Stepped terraces in the Ghagghar valley, east of Chandigarh. The tear fault has offset the HFT Fault along with the frontal Siwālik Range. **a** and **b** After Philip et al. (2014) **c** After Philip et al. (2011) **d** After Malik et al. (2008)

Plate 3.1 The Ghagghar cuts its way through the Siwalik terrain forming erosional stepped terraces. *Inset* Entrenched channel of the Ghagghar (From Anonymous 2015)

Valley (Nakata 1972) implies that the entire expanse of the land east of the Saraswatī course has been in a tectonic ferment in the Late Holocene time. Geodetic measurements show that the Dūn Valley is rising at the rate of 1 mm/yr (Rajal et al. 1986).

The northern part of the Siwālik terrane (Fig. 3.3) has witnessed far stronger movement. In the Pāontā Sāhab belt the Lower Siwālik sandstones have moved southwards along a thrust and are now resting on the Late Pleistocene to Holocene gravels (Joshi et al. 2005). In the Tons valley the Nāhan Thrust has placed the Eocene Subathu rocks upon a scree deposit containing 38,270 ± 2480 year old carbonaceous matter (Krishnaswamy et al. 1970).

To the west in the Mārkandā valley (through which the Tamasā branch of the Saraswatī once entered on to the plains), the terrain (Figs. 2.2, 3.3 and 3.4) is tectonically far more resurgent. The Mārkandā valley is carved along the zone of the fault that tore apart the Siwālik frontal range and pushed the dismembered western block northwards relative to the eastern (Fig. 3.4). Palaeoseismic investigation through systematic trenching has revealed two large rupture planes related to the earthquakes in A.D. 1294 and A.D. 1423 and the third one in A.D. 260 (Kumar et al. 2001). These earthquakes resulted from slip of the order of 4.6–2.4 m. In the past, the rupturing of the ground lifted 27 m up the fluvial terraces, dated at 4896 ± 68 yr to 5669 ± 205 yr. The earthquakes originating in these rupture planes occurred in 3349 ± 55 yr B.P. and 1523 ± 99 yr B.P.—the latter one resulting from 9 m slip (Kumar 2001). This event that occurred in 1523 ± 99 yr B. P. must have been of great magnitude. Yet another study of similar kind demonstrated repeated rupturing of the ground and giving rise to a 15–38 m high cliff. The rupturing caused earthquakes in A.D. 1300 and 1400 (Malik et al. 2003). Palaeoseismic studies have revealed that an earthquake of magnitude more than or equal to 7.5 occurred 29,300 years ago and another of M 7.2 happened 5800 years B.P. (Philip et al. 2012).

Further west, in the Ghagghar valley belt the Pinjaur Garden (Figs. 2.2, 3.3 and 3.4) was rocked by an earthquake about 4000 yr B.P. This event caused 2 m displacement of land over a length of 45 km and gave rise to 2–25-m high WNW–ESE trending escarpment (Malik and Mathews 2005; Malik et al. 2003). In the Ghagghar channel the stream strath was uplifted and upwarped, and the alluvial deposit tilted and truncated (Singh et al. 2011).

The Ghagghar has cut its channel along a NW–SE oriented (Figs. 3.3 and 3.4) fault that has displaced the Himālayan Frontal Fault along with the frontal Siwālik hill range. The tear fault has also uplifted recent terraces on which the townships of Kālkā and Pinjaur are located and also those on which the Kaushalya and Ghagghar rivulets flow (Mallik et al. 2003). A 15–18 m high scarp formed by faulting up of the ground is related to the earthquake which occurred in the interval A.D. 1300–1400 (Malik et al. 2003).

Further west, the Satluj River—the Shatadru branch of the Saraswatī—has cut 950 m deep straight channel across the frontal Siwālik hill, the river course coinciding with the NE–SW oriented westward inclined tear fault that passes by Ropar and dislocating the Siwālik hill (Singh et al. 2008).

West of the Satluj valley, the younger sediments deposited in the Soan valley are overridden by the Middle Siwālik along the Satlatta Thrust—an intrabasinal thrust within the Siwālik terrane. The thrusting has displaced a gravelly terrace west of the Pong Dam and uplifted a recent deposit 140–160 m above the ground (Talukdar and Sudhakar 1977). The continuing growth of the anticline seen in the Surinsar area is a pointer to the on-going movement in this tectonically disturbed belt. Recent works demonstrate that the anticlinal ridges of Chandigarh and Jaunauri (WNW of Chandigarh) are growing laterally, that is sideways. This is borne out from folds presently developing due to gradual movements on active faults, pop-up structures, occurrence of mantle of alluvial fans on the hinges of the folds, stream piracy and drainage pattern (Decaillau 2006). The shortening of the crust due to very strong northward compression is also absorbed by the transverse faults through which the rivers Ghagghar, Satluj and Beas have found their ways out on to the Panjab Plain. The faults at Kala Amb, Ghagghar, Ropar and other places are thus transferring the ongoing stresses and resultant deformation.

It is more than obvious that this belt in which the Tons, the Markanda, the Ghagghar and the Satluj have carved their courses was repeatedly convulsed during strong earthquakes, and witnessed spasms of tectonic turmoil. It is therefore understandable that there were spectacular drainage deflections, channel changes by rivers, and beheading and piracy of streams. It seems that under the cover of the Saraswatī Basin the crust is restlessly adjusting itself to gathering stresses and strains generated by northward moving Indian plate.

Tectonics of Hidden Ranges in the Saraswatī Floodplain

Geophysical investigations in western Indo-Gangetic Plains reveal existence of NW–SE to WNW–ESE oriented subsurface hidden the *Lahore–Sargodhā Ridge* branching off from the Arāvali orogenic belt (Fig. 3.5). In the Kirānā area in Pākistāni Panjāb it is represented by inliers (hills and hillocks) of Precambrian rocks

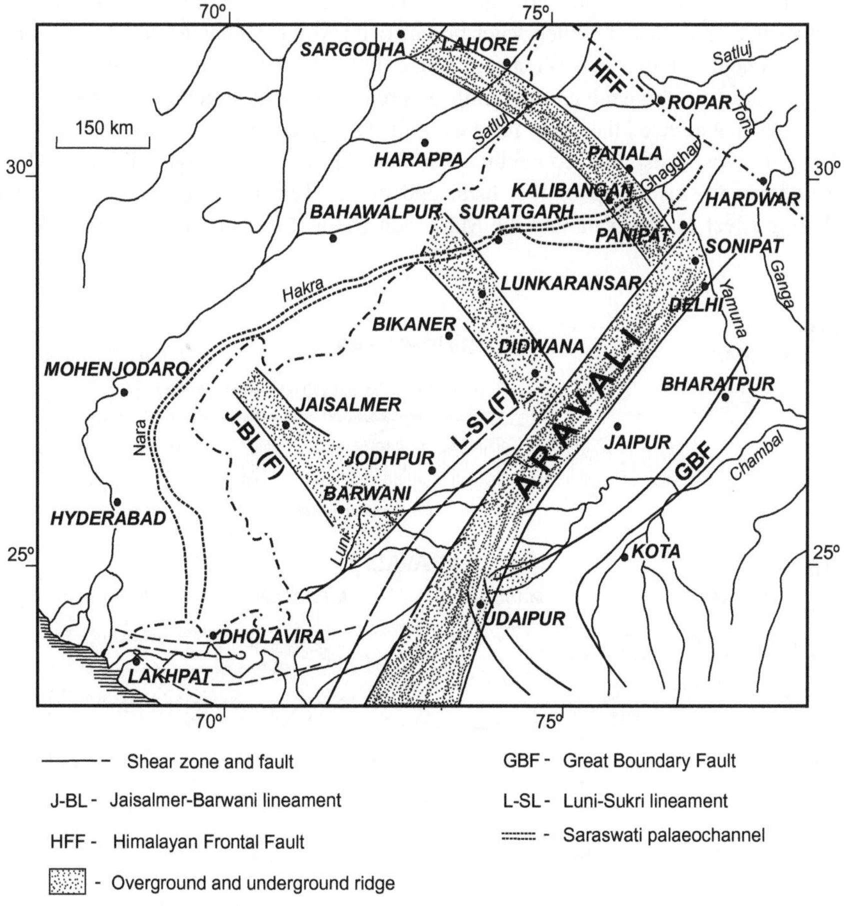

Fig. 3.5 Diagrammatic sketch showing hidden ridges under the thick cover of the alluvial–fluvial sediments of the Saraswatī Basin—in the Panjāb Plain (Based on Sinha-Roy et al. 1998; Roy 1999 and Biswas 1987—From Valdiya 2013)

(Fig. 3.5). This subsurface ridge is believed to have controlled the tectonic framework of the vast Haryānā–Panjāb Plain made up of Late Pleistocene to Holocene alluvial–fluvial sediments (Thussu 2006). The distribution of epicentres of earthquakes in the period 2001–2004 not only conform clearly to the trend of the Lāhore–Sargodhā Ridge, but also demonstrates that it is cut along its length by faults that are active. The A.D. 1437 earthquake of M.6.0 and the A.D. 1827 event at Lahore seem to be related to the tectonic movements in the under-surface WNW trending ridge (Thussu 1999). Nearly 75 % of earthquakes were clustered in the NW–SE direction of the hidden ridge (Shukla et al. 2007). Fault-plane mechanisms indicate dominantly thrust movement, with a small strike-slip component (Bansal and Verma 2012). The 5th March 2012 earthquake of M 4.5 occurred near the surface expression of the Mahendragarh–Delhi Fault located NW of Delhi.

In the Pehowā area a number of water bodies are seen located at the distal end of the zone of Terminal Fan (Mukherjee 1976). The central part of the Choawālā is naturally dammed—is ponded—giving rise to a linear Bhupinder Sāgar lake (Srivastava et al. 2013). Other water bodies occupy palaeochannels or depressions. The drainage in this part is often segmented and exhibit anastomosing pattern. The segmentation of drainage and ponding of streams are consequences of neotectonic activities that affected the upper reaches of the Ghagghar domain. Neotectonic uplift along the east–west line coinciding with the Gurla-Ladwa high (Thussu 2006) resulted in the rising up of the distal end of the Terminal Fan, and consequent adverse effect on the Nala and the Mārkandā River (Mukherji 1976) (Fig. 3.5).

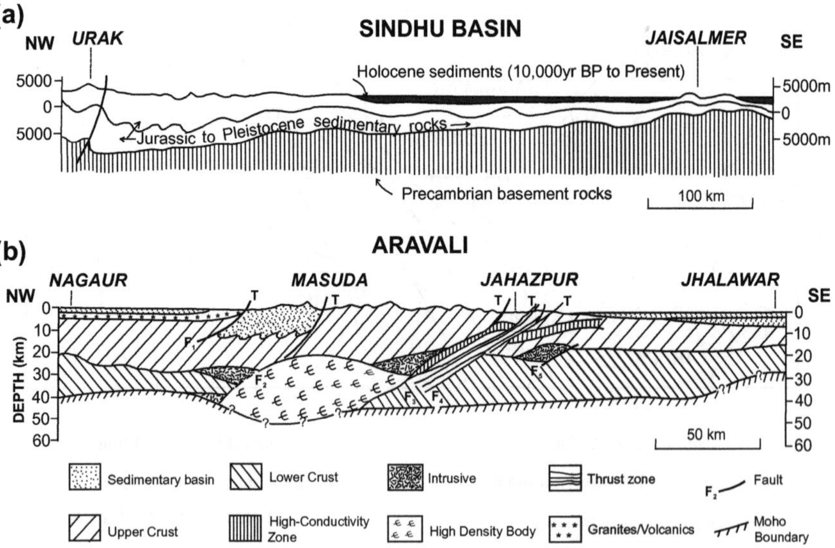

Fig. 3.6 The stratigraphic layout and the structural design of the Saraswatī domain in Rājasthān and adjoining Pākistān under the thick column of fluvial–aeolion sediments **a** Valdiya (1999b) **b** Tewari et al. (2000)

Southwest of this high, the lower layers of the Kālibangan sedimentary column shows marked development of cleavage, developed presumably due to ground shaking. This is borne out by the collapse of massive wall of the fortification (Bisht 2000). The earthquake is dated to 4700 yr B.P. There are quite many clear evidence for post-Neogene tectonic movements along the Arāvali Range (Fig. 3.7) in Rājasthān (Sen and Sen 1983).

Further southwest, two hidden ridges have been recognized—*the Didwānā–Lunkaransar Upwarp* bound by shear zones or faults and the *Barwāni–Jaisalmer Ridge* demarcated by prominent faults (Figs. 3.5 and 3.7). The one related to the Barwāni–Jaisalmer Ridge is 1000 km long and believed to be responsible for the earthquakes of 1907, 1959 and 1993 (Krishna Brahman 1993). The earthquakes gave rise to peculiar circular features in the desert terrain (Ramasamy 1999). The November 1991 Jaisalmer earthquake of M 5.6 resulted from the reactivation of the NNW–SSE oriented subsurface *Barmer–Konai Fault* (Joshi 1997). On the other hand the earthquakes of 1891 and 1976 are attributed to movements on the NE–SW trending *Luni–Sukri Fault* (Fig. 3.5). This 750-km fault extending from the Great Rann of Kachchh to Sambhar defines the boundary of the Arāvali Mobile Belt (Sinha Roy 1986; Sinha Roy et al. 1998). In Churu district the N–S *Sardārshahar*

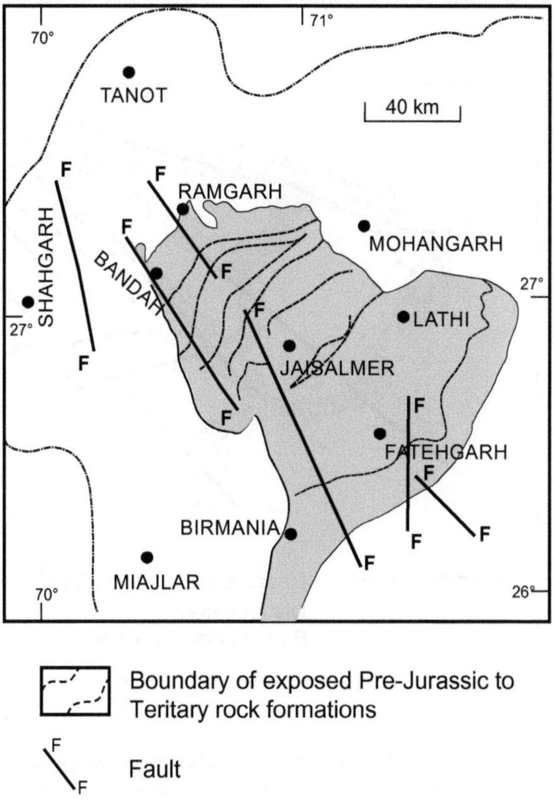

Fig. 3.7 Faults in the Jaisalmer region in western Rājāsthan that is characterized by subsurface grabens (Based on Singh 1996; Sinha-Roy et al. 1998)

Fault is associated with a depression filled with riverine sediments (Ahmad 1986). This fault is manifest in the development of step-like landforms, the steps descending westward.

Underneath the realm of the desert, a number of Neogene depressions have been identified on the basis of data obtained through field work, dug wells and interpretation of subsurface data. West of the *Sardārshahar Fault* there is a series of linear grabens and horsts resulting from sinking and uplift along step faults, the fault blocks downthrown westwards (Wadhawan 2015). The Sardārshahar Fault is traceable west of Churu (Fig. 3.7). These grabens are filled with Quaternary sediments—fluvial, lacustrine and aeolian. Since the fractured framework of the desert region has witnessed neotectonic resurgence (Wadhwan 2015), it is but natural that landforms were modified and rivers changed their courses. It is likely that reactivation of the faults of these grabens may have influenced the course of the Saraswatī and its many palaeochannels.

Interpretation of satellite imagery and aerial photos, coupled with drilling in connection with oil exploration in the Jaisalmer region, demonstrate that the desert terrain in Rājasthān is riddled with multiplicity of long deep NE–SW and ENE–WSW faults (Bakliwal and Wadhawan 2003; Wadhawan 1990) that are continuously reactivating as borne out from straight courses of rivers and streams and by alignment of many major and minor palaeochannels (Fig. 3.8) along these trends

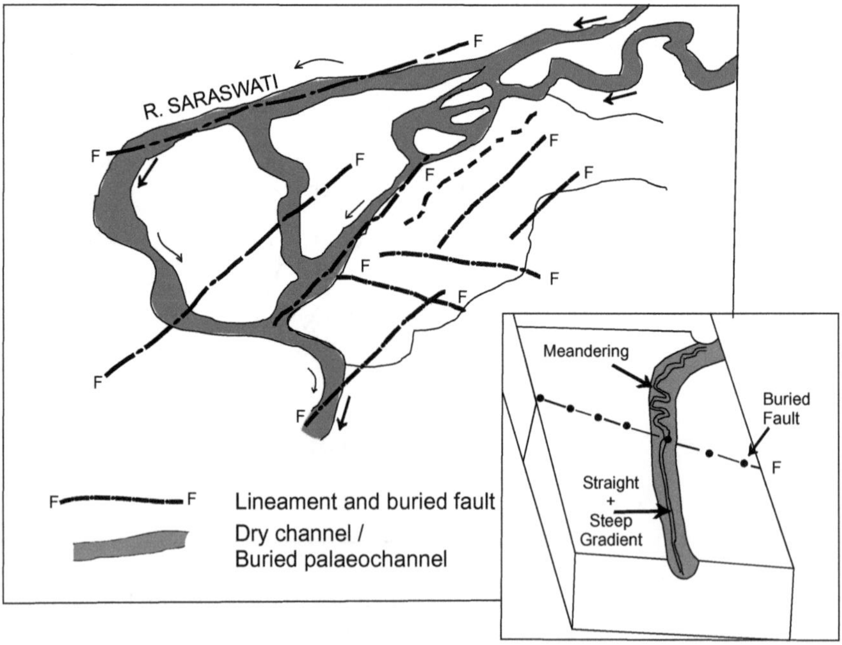

Fig. 3.8 Faults affecting the lower reaches of the Ghagghar–Hākrā River and its many channels in northwestern Rājasthān. Note the faults controlling the drainage pattern (After Mitra and Bhadu 2012)

(Mitra and Bhadu 2012). The Saraswatī channels were thus controlled by these faults which caused deformation in the subsurface rocks as confirmed by drilling. There are faults exhibiting evidence for strike-slip (sideways) and dip-lip (up-and-down) movements to various extent. The cumulative effect of fault movements is the uplift and sinking and, in some areas horizontal displacement of the ground and resultant modification of landforms.

Under these circumstances the rivers and streams were forced to change their courses, sometimes gradually sometimes abruptly, as the pattern of palaeochannels in the Luni River and the Ghagghar (Saraswatī) demonstrate unambiguously (Kar and Ghose 1984; Yashpal et al. 1980; Bakliwal and Grover 1988; Ghose et al. 1979; Kar 1988, 1999a, b; Sahai 1999; Gupta et al. 2004, 2008).

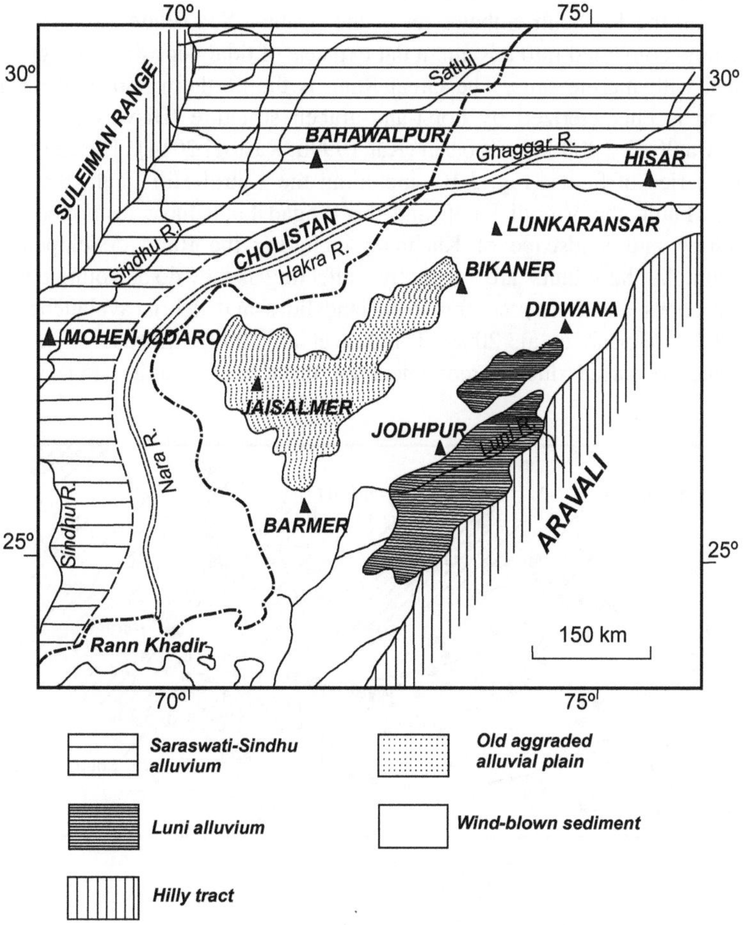

Fig. 3.9 Geological map of the middle and lower reaches of the Saraswatī (presently Ghagghar–Hākrā–Nārā) River showing exposed older rocks amidst the desert sands and alluvial sediments (Based partly on Raghav 1999)

Beyond the hidden Barwāni–Jaisalmer Upwarp, the NW–SE trending *Jaisalmer–Mari Arc* passes through Shāhpur. At this place the meandering nature of the river abruptly ends, and onwards it flows in a linear course as if structurally controlled (Mitra and Bhadu 2012). More important is the fact that the gradient of the river bed is steeper, implying uplift of the faulted block through which the river flows. Drilling has established that the faults recognized on the surface extends to considerable depth (Fig. 3.8).

Seismic Upheavals in the Saraswatī Delta

The terminal reaches of the Saraswatī was represented by the Nārā, which discharges into the Kachchh embayment of the Arabian Sea (Plate 2.6). The embayment is now converted into a Rann, a flat expanse of tidal deposits impregnated and encrusted with marine salts. The Great Rann is the northern part of the Kachchh Basin and is characterized by horst-and-graben structure (Biswas 1987). It was originally a 300 by 200 km rift basin (Kar 1999b; Rastogi 2014). The roughly E–W trending series of faults (Fig. 3.10) including the Katrol Hill Fault, the Kachchh Mainland Fault, the Vigodi Fault, and the Wagad Fault have fashioned the geomorphology and landscape of Kachchh, including the Rann developed over a depression. All these faults are very active, affecting as they do as young deposits as colluvial cones and fans resulting from landslides and debris avalanches (Malik et al. 2008; Morino et al. 2008; Thakkar et al. 1999). Moreover, quite many earthquakes occurred due to movement on these active faults (Fig. 3.10). The

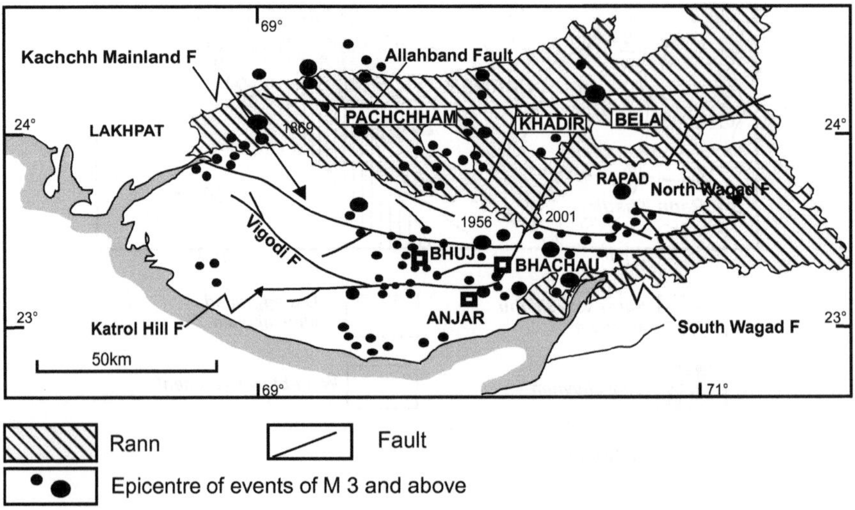

Fig. 3.10 Kachchh Basin is cut by many roughly E–W oriented faults that are related to horst-and-graben structure (Based on Biswas 1987; Rastogi 2014)

northern fault is responsible for evolution of the 80 km long, 16 km wide and 3–6 m high E–W trending *Allāhband Ridge* (Fig. 2.6). Palaeoseismites have been unearthed which testify to earthquakes occurring 3000 yr B.P. and 1000 yr B.P. in the vicinity of the *Allāhband Fault* (Rajendran and Rajendran 2000). The Harappan port town of Dholāvīrā is located in the islet of Khadīr in the proximity of the Allāhband Fault. This town was rocked by earthquakes in 4500 yr B.P., 4200 yr B. P. and 2200 yr B.P., as deformation of walls and collapse of walls bear out (Bisht 1991, 1997, 2013; Joshi and Bisht 1994). It was the movement on the Allāhband Fault that generated A.D. 1819 earthquake of magnitude M_w 7.5. The rise of the Allāhband Ridge obstructed for seven years until 1826 the flow of the Nārā River. Concomitantly, two depressions formed and Fort Sindri sank into one of these and Fort Basta was destroyed following ground subsidence (Thakkar et al. 2012).

Not far to the west of the Nārā (Saraswatī) delta, the Karāchī–Makrān coastal belt is tectonically and seismically a resurgent terrane. Apart from being rocked repeatedly by large earthquakes, this belt has been rising up fast. It has risen up at least nine times in the Holocene epoch alone, as borne out by as many levels of beach terraces (Fig. 3.11) along the Balochistān Coast (Page et al. 1979). The uplift must have caused decrease in the gradient of rivers, including the Sindhu. This would have resulted in ponding of the rivers and formation of upstream lakes. The lake waters must have submerged the floodplain, forcing the people to flee (Raikes 1964; Flam 1993).

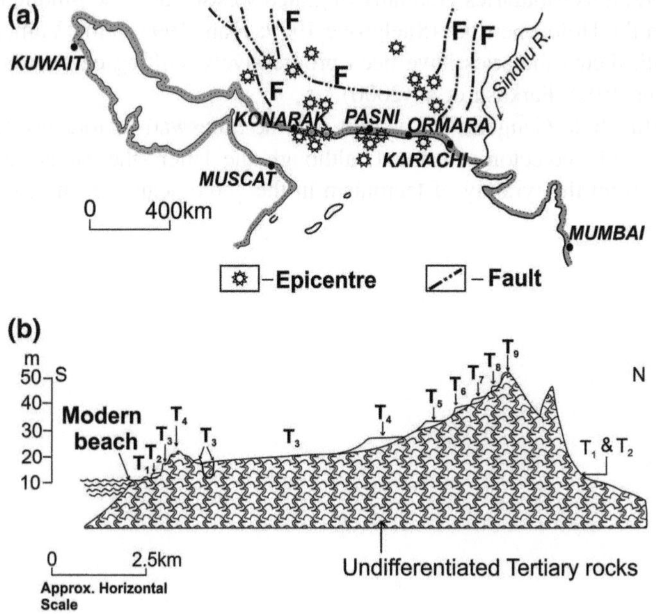

Fig. 3.11 **a** Major active and seismogenic faults and earthquake epicentres in Balochistān coastal belt. **b** Nine levels of beach terraces at different elevations along the Makrān coast in Balochistān indicate as many pulses of uplift in the last 10,000 years (From Page et al. 1979)

Saraswatī Basin: A Land of Recurrent Tectonic Upheaval

It would be obvious from the preceding discussion that the vast expanse of the land traversed by the Saraswatī and its tributaries has more than once experienced tectonic convulsion. It was severely shaken by earthquakes of large magnitude and the ground was displaced. Some parts of it rose up and other parts sank along the faults that dissect it. Geomorphological and riverine evidence point to the high intensity of neotectonism that affected the belts where the two branches of this river descended on to the Indo-Gangetic Plains, and in the area where it emptied itself into the sea. Flowing through the land repeatedly convulsed by crustal upheaval, the Saraswatī indeed had a very tempestuous life as described long, long ago in the *Rigved—apasām apastamā* (6.61.13).

Under the circumstances when the land rose up in some parts, sank in others, and the ground tilted one way or the other, it is quite natural that the rivers and streams deviated, shifted, swung and swerved from their original courses, flowed through more than one channel, were ponded behind impediments created in their pathways due to uplift, and were beheaded and robbed of their waters by other rivers. This indeed happened to the Saraswatī and its branches that followed a path riddled with active faults. The changes that took place considerably modified the landform so much so that it is not the same today as it was when the Saraswatī flowed during the prime of its life.

These developments are not unique to the Saraswatī River. While the Sindhu (Indus) and all its tributaries gradually migrated westwards—the Sindhu as much as 160 km in the Holocene time (Snelgrove 1979; Flam 1993)—the Yamunā and the Gangā with their tributaries have been progressively shifting eastwards (Bakliwal and Sharma 1980; Parkash et al. 2000).

The entire Indo-Gangetic Plains, not just the Saraswatī sector, has experienced the impacts of neotectonic ferment, although the latter (the Saraswatī) suffered drastically from the tyranny of tectonism in the geologically recent time.

Chapter 4
Portrayal, Identification and Delineation of the River Saraswatī

Poetic Portrayal in *Rigved*

The Saraswatī was a highly venerated river in the times it watered the vast expanse of the land known as *Saptasindhav* (Fig. 4.1)—the land of seven rivers (*Rigved*, 8.24.27) namely Saraswatī, Parushnī, Vipāsh, Asiknī, Vitastā, Sindhu and Kumbhā. Presently known as the Ghagghar, the Beās, the Rāvi, the Chenāb, the Jhelam, the Indus and the Kābul rivers respectively, these seven rivers held sway over the larger part of northwestern India.

The *Rigved* describes the river Saraswatī in glowing terms. It was *naditame*, the best of all rivers of the land (*Rigved*, 2.41.16) surpassing all other rivers in splendour and benevolence *mahimnā mahinār* (*Rigved*, 6.61.13). It was a mighty river *maho arnah* (*Rigved*, 1.3.12), abounding in water that was pure from the source in the mountains to the sea *Shuchiryati giribhyā ā-samudrāt* (*Rigved*, 7.95.2). Flowing swiftly (*prasare*), it was most impestuous of all rivers *apasām apastamā* (*Rigved*, 6.61.13) that descended with roar (*charati roruvat*) down the slope (*Rigved*, 6.61.8), its fierce (*ghor*) currents gurgling (*Rigved*, 6.61.7) through its canyon course.

The *Rigved* thus presents a panegyrical account of the river revered by the people of *Saptasindhav* for whom it was a lifeline. It was the most bountiful river, the finest of all the rivers of the land in which the composers of *Rigved* lived.

How important the Saraswatī was to the people of the Vedic time would be evident from the fact that practically every chapter (*mandal*) of the *Rigved* mentions it: There are 75 *shlokas* or verses in this great work mentioning the Saraswatī. The *Yajurved* mentions the Saraswatī 68 times.

K.S. Valdiya, *Prehistoric River Saraswati, Western India*,
Society of Earth Scientists Series, DOI 10.1007/978-3-319-44224-2_4

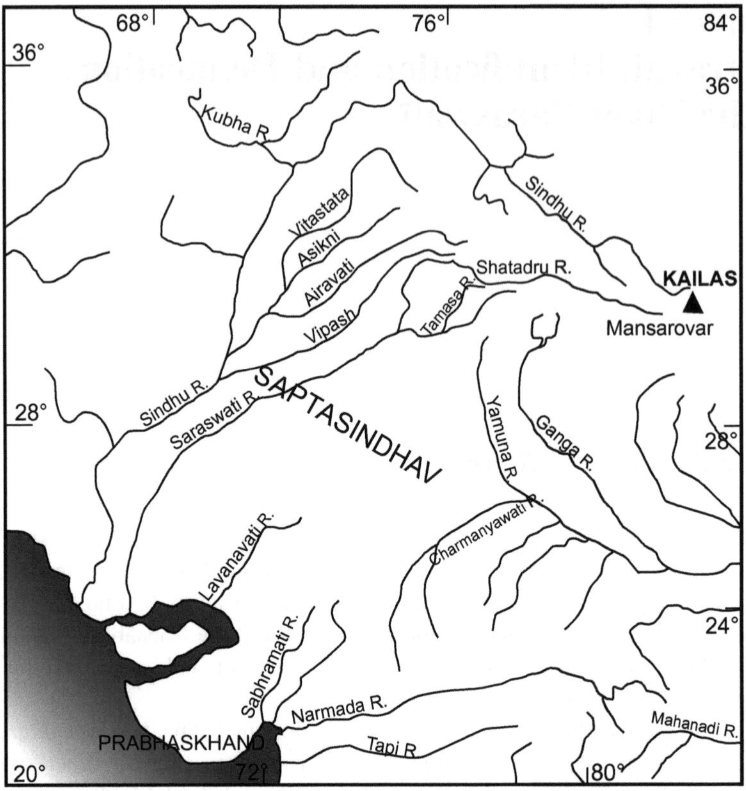

Fig. 4.1 The land of seven rivers, encompassing the floodplain of the Sindhu and the Saraswatī river systems, was known as *Saptasindhav* in the Vedic time (After Valdiya 2012)

Description in Purāns and Epics

In *Skand Purān* the Saraswatī is shown as originating in the snowy Himālaya and flowing down to the foothills covered with *plaksha* (horse chestnut) trees—*Himvant girim prāpya plakshāntatah vinirgata* (*Skand Purān*, Prabhaskhand (7), Part 33, 49).

In its way down to the plains, the river cuts through multitudes of mountain ranges (*saisha shailsaharāni vidāryam*) and flowed with great force (*bheemvegam*) at the spot where sage Vasishtha's hermitage was located *aashramo ve Vasishthasya sthanutirth a bhavanmahaas* (5) (*Mahābhārat*, Shalya Parv, 41, 4,5).

Further downstream, the Saraswatī broke into many "*dhārās*" (tenors) or channels, including the ones known as Apagā, Mandākinī, Madhustravā, Vāsu, Kaushikī, Drishadwatī and Hiranyawatī. These channels of the river were seasonal, for water flowed through them only during rainy season—*varshākālavahā sarvā varjyitvā Saraswatīm* (*Vāman Purān*, 34, 6,7,8).

This statement is very significant. While in the *Rigved* time the Saraswatī was a mighty river full of water, by the time the *Vāman Purān* was composed, quite many of its multiple channels had lost water and become seasonal, implying decline in the discharge in the main river.

After watering the Kurukshetra region, the Saraswatī turned westwards, and flowed through forests such as Sitāvan, Aditivan, Kāmyakvan, Dvaitvan, Vyāsvan, Phalakivan, Suryavan, Madhuvan and Sheetvan (Fig. 1.4). Some of these sanctuaries hosted *āshrams* (schools of learning) of the celebrated sages (*rishis*)—*nadi pravāhsamyukta Kurukshetram viveshah, tatrā sā rantukam prāpya punyatoyā Saraswati, Kurukshetram samāplavya prāyātā paschimām disham; Kāmyakam cha vanam punyam tatā Adityavanam maha; Vyāsasya cha vanam punyam Phalakivanameva ch; tatra Suryavanasthanam tathā Madhuvanam mahat; punyam Sheetavanām nām sarvakalmashnāsnam* (*Vāman Purān*, 33, 1,2,4,5).

Obviously, west of Kurukshetra the floodplain of the Saraswatī was a forested tract, as is common with the rivers abounding in water round the year. When Balarām, the elder brother of Krishna, was on a pilgrimage along the course of the Saraswatī from its mouth at the sea (around Prabhāskshetra) to its headwaters, he came across a pilgrimage place *Nāgdhanwā* swarming with snakes (*Mahābhārat,* Shalya Parv 37). It must have been a waterlogged terrain with many swamps and pools as are associated with the rivers that flow sluggishly, in their meandering courses (*gātva chaiva—Nāgadhanvānam tirthamāgamagamadchutah; yatra pannagarājasya Vāsukeh sanniveshnam*) (*Mahābhārat*, Shalya Parv 37, 30).

Further downstream, the situation was quite different. The Pāndavs during their exile (*vanvās*) and later Balarām in his trek along the Saraswatī, noticed that after the plough-shaped bend at a place Vinashan, the river disappeared under the mass of sands (*tato Vinashanam rājan jagāmāth halāyudhah; yatra nashtā Saraswatī* (*Mahābhārat,* Van Parv 25, 1). Still further downstream, the Saraswatī reappeared in a number of spots such as Chamsodbhed and Udapān—both the centres of pilgrimage. Balarām was able to recognize the underground (*antahsalila*) currents by noticing moisture in the soil and the greenery that carpeted the surface *Udapnamgachhatvaraav* … (89) snigdhtavaadoshdheenam cha bhumeshch … jaananti siddhaa … nashaamapi Saraswateem (90) (*Mahābhārat*, Van Parv 35, 1, 89, 90).

Although the Saraswatī had lost its water at this place, the phrase *Saraswatybdhīsangam* in the *Shalya Parv* 33, 70 in the *Mahābhārat* indicates that it did flow into the sea at least during the rains (Chakraborti 1982).

Balarām had started his journey from the place where the Saraswatī met the sea (*samudram paschimam gatwa Saraswatyabdhisangamam*). (*Mahābhārat*, Shalya Parv 35, 77).

It is quite obvious from the narratives in the ancient texts that while the *Rigved* portrays the Saraswatī as a Himālayan-born mighty river abounding in water and emptying itself in the sea, the later-day *Skand Purān* and *Vāman Purān* describe it breaking into a multiplicity of channels, some of which were just seasonal through which only rain waters flowed. On the other hand, the epic *Mahābhārat* paints a dismal picture of the Saraswatī that it had degenerated in its *middle reaches* and gone under the pile of sands in the *lower reaches*. In the sandy desert, the Saraswatī had become subterranean—an underground (*antahsalila*) river.

The *Rigved* is earliest amongst the old texts. The *Purāns* were composed sometimes before the great war of Mahābhārat, which happened 3500 yr B.P., and the epic *Mahābhārat* was written after this debilitating war (Valdiya 2012; Bhatnagar 2014).

It is interesting to note that in most of the *Purāns*, there is no mention of Yādav hero Krishna after his leaving Mathurā to settle down at Dwārkā on the sea coast. Indeed there is no mention in the *Purāns* of the *Mahābhārat War* and of the happenings thereafter. It seems that the Krishna Dwaipāyan 'Vyās' composed the *Purān*(*s*) sometime after Krishna left Mathurā and before the great civil war (Valdiya 2012). On the other hand, the epic *Mahābhārat* embodies comprehensive accounts of this war and of the developments after that event. This implies that the epic was written after the *Mahābhārat War*. The War happened 1478 yr BCE (3478 yr B.P.) according to Iyengar (2003) and 1500 yr BCE (3500 yr B.P.) in the opinion of Bhatnagar (2014). These dates are deduced on the basis of analysis of planetarium software related to solar and lunar eclipses and positions of stars and planets.

For three major reasons I am inclined to accept the time of the Mahābhārat War sometime between 3500 and 4000 yr B.P. *One*: The 20-m uplift of the stream terrace in the Yamunā valley related to reactivation of a major fault that tore apart the Siwālik Range and diverted the eastern branch of the Sarawatī (*Tamasā* River) occurred sometime between 3878 and 3448 yr B.P. (Wesnousky, et al. 1999) resulting in drastic decline of the Saraswatī discharge. *Two*: As a consequence of decline in flow, there was a great exodus of the Harappans from the middle reaches around 3750 yr B.P. (Thapar 1975). *Three*: The archaeologists believe that the Painted Grey Ware Culture (3200–2500 yr B.P.)—which represented a agrarian economy, iron technology and growing specialization in craft (Bhan 2000)—co-incides with the time span described in the epic *Mahābhārat* (Lal 1998, 2002; Joshi 2008).

It is therefore surmised that the epic *Mahābhārat* was written sometime later than 3500 yr B.P., and the *Purān* before that time (Valdiya 2012).

Together, the *Rigved*, the *Purāns* and *Mahābhārat* transmit across a time span of over three thousand years a remarkably accurate descriptions of a great river that was the lifeline of the people. The narratives also chronicle clearly the three phases in the life of this river—its youth, its old age, and its demise. The extraordinarily wide sand-filled dry channel of the rivulet called Sarsuti in the northern Haryānā, Ghagghar in the larger part of this state, Hākrā in Cholistān (Pākistān), and Nārā in Sind (Pākistān) represents the river once known as the Saraswatī.

It may not be out of place to dwell on the historicity of the *Rigved*, the *Purāns* and epics. Comparing the descriptions the geographic layout and of the flora and fauna in the *Purāns* and the epics (*Mahābhārat* and *Vālmiki Rāmāyan*) with the landforms extant during the Later Harappan period, it becomes clear that the Harappans and the people of the Purān times lived in the same land, with quite similar the natural and socio-cultural environments (Rao 1991, 1997; Mishra 1993; Gupta 1996; Lal 1997, 2002, 2005). From archaeological evidence it seems that the Harappan Civilization cannot be contrasted or separated from the Vedic Culture (Bisht 1998). Not only is there a very close similarity but also the two are contemporaneous (Singh 2008).

However, quite a few historians view the two cultures as belonging to two different periods, and hold the view strongly that the Vedic Culture cannot be compared with the Harappan Civilization, for the social structure and the economic conditions of the two were quite different (Sharma 1999; Thapar 2006; Kochhar 2000; Bhan 2000).

Taking in totality the ample evidence yielded by archaeological investigations, it appears that the Later Vedic Culture resembles the Mature Harappa Civilization 5500–3900 yr B.P., and the Mahābhārat Culture is comparable to the Painted Grey Ware Culture 3200–2500 yr B.P., ironically, while there is literature of the people of the Vedic time, there is no archaeological evidence, and while the Harappan Civilization left profuse archaeological evidence, they have no literature to construct their history. I believe that the ancient literature, as already adumbrated, describes the Harappan Civilization, albeit in a verbose language characterized by superfluity, metaphors, and allegories (Valdiya 2012). Moreover, the authors of the Purānic literature adopted the mode of story-telling, presumably to convey the subjects to the general public in a memorable and enjoyable manner.

Routes of Trade and Invasion in Historical Times

Despite the River Saraswatī degenerating into an extraordinarily wide dry channel with scarce water, its floodplain remained attractive all through the historical times to explorers, traders, invaders and marauders. Emperor Harsh Vardhan in the early seventh century BCE, performed the last rites of his father on the bank of the Prāchi Saraswatī that then flowed not far from Kurukshetra (Fig. 4.2) as described by Bān Bhatt in the biography *Harsh Charitra* he wrote (Danino 2010). Alexander the Great unsuccessfully tried to penetrate the western fringe of the Saraswatī land in 326 BCE. As already stated earlier, from Mahmūd Ghaznavi (997–1025 CE), Mohammad Ghori and Taimūr of Samarkand (1398 CE), Bābar (1526 CE) to Ahmad Shāh Abdāli (1761 CE) invader after invader plundered the land as they marched towards Delhi (Fig. 4.2). The well-known traveller and writer Ibu Batutā (1325–1354 CE) was fascinated by the fields of rice and sugarcane in the floodplain of the anomalously wide channel of river that traversed the land from east to west. Prosperous and bustling trade towns like Mārot, Sirsā and Hānsi were situated on the road travelled by explorers, traders and invaders. Fort "Sarsuti" near Sirsā stands high as a mute sentinel of the river that once flowed past this centre of commerce.

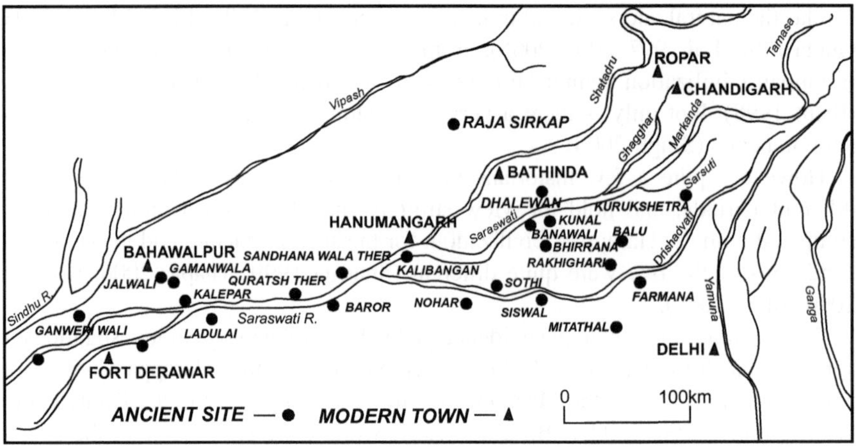

Fig. 4.2 Once bustling trade cities on the banks of a disproportionately wide (4–10 km) channel of a nearly waterless course now called Ghagghar in Haryānā and Hākrā in Cholistān (After Lal 2002)

Delineation by Surveyors and Geographers

The builders of roads between Sirsā and Bahāwalpur across the dreary waste of the Thār Desert noticed that the banks of an extraordinarily wide waterless channel filled with sands were dotted with a large number of ruins of developed settlements, some of them of the dimension of towns (Mackeson 1844). Moreover, on digging wells they were rewarded with abundant sweet (potable) water in the desolation of the desert. According to Michel Danino (*The Pioneer*, 20 October 2012) the *Library Atlas* published in 1760 A.D. by Bryce, Collier and Schmitz shows a river *Soorsutī* (Sarsutī) joining the *Guggur* (Ghagghar) in what was then Panjab. The then Surveyor-General James Rennel in his *Map of Hindoostan* (1788) showed a dotted line marking the wide dry channel as the course of a river under the caption "Gaggar", "Hankra" and "Sursooti" (≡Ghagghar, Hākrā and Sarsutī).

de Saint-Martin (1860) believed that the streams known as Mārkāndā, Sarsutī, Ghagghar and Chautang in Haryānā were the main tributaries of the Vedic Saraswatī that flowed through the waterless tract (Haryānā–Rājasthān) to the Gulf of Kotdi (Kachchh). Raverty (1892), who surveyed the wide swathe of the land encompassing Sindh and Kachchh, asserted that the Sarasutī, the main tributary of the Ghagghar, flowed into the Rann of Kachchh through the Eastern Nārā channel. The Rann according to him, was a navigable arm of the sea.

The 1908-*Imperial Gazetteer* described the Satluj River flowing into the Hākrā in 1000 CE and also later around 1245 CE. Thereafter it took northwesterly course resulting in the drying up of the Hākrā. Still later the Satluj returned to join the Ghagghar, and finally abandoned it in 1796 CE and met the Beās.

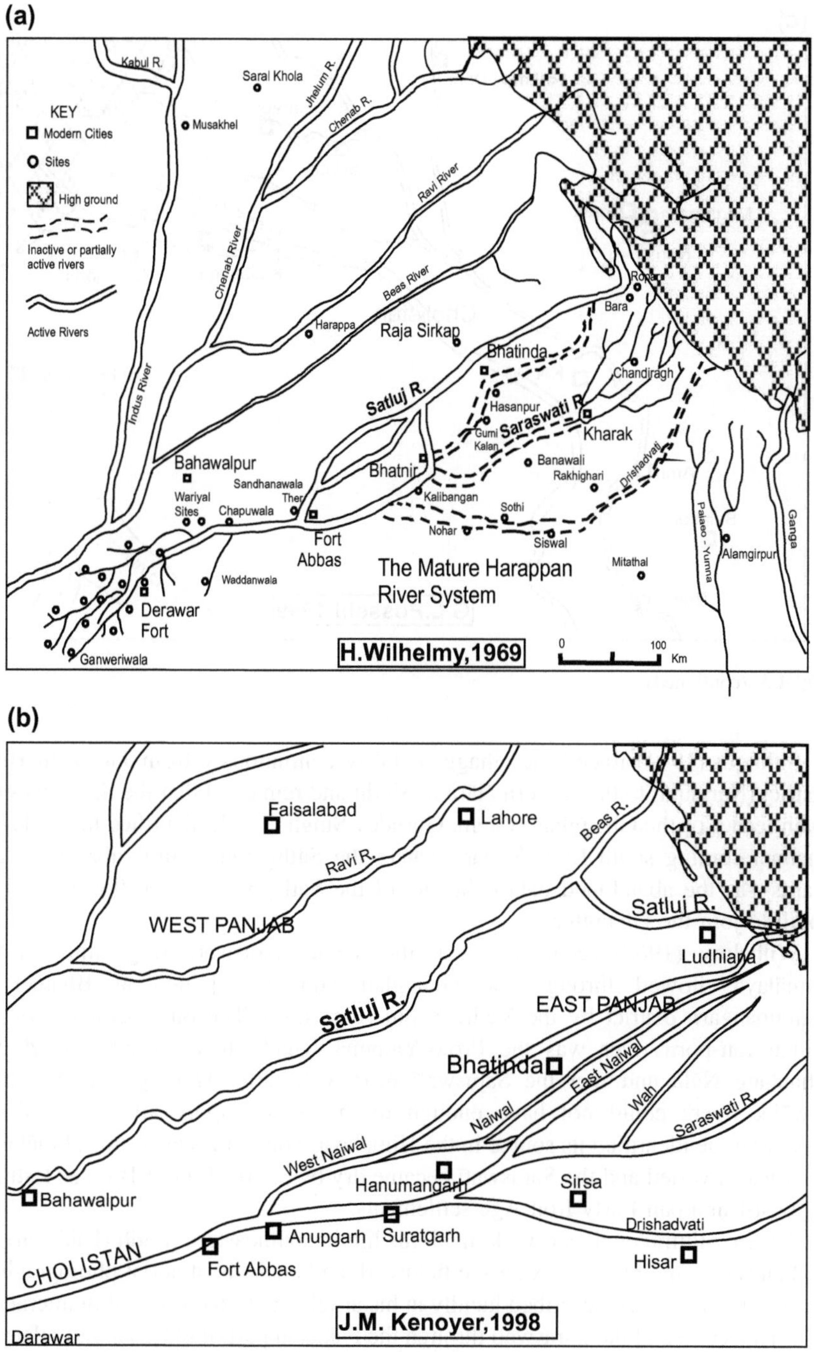

Fig. 4.3 The Saraswati as depicted by **a** Wilhelmy (1969), **b** Keynoyer (1998), and **c** Possehl (1999) (Based on Danino 2010)

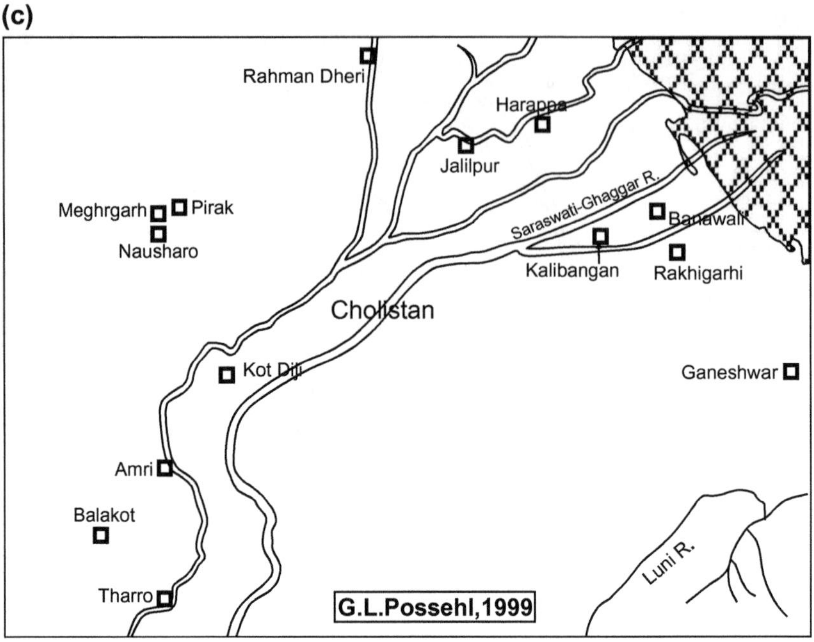

Fig. 4.3 (continued)

Siddiqui (1944) traced the Ghagghar fairly continuously from Jakhal in Hisār district (Haryānā) to the Eastern Nārā in Sindh and remarked that the Satluj was the main feeder of the Ghagghar system. Gurudev Singh (1952) identified the wide dry channel coming south from Ropar—where the Satluj today turns abruptly westwards—as the abandoned palaeochannel of the Satluj which then flowed straight south to join the Ghagghar.

Wilhelmy (1969) recognized that the Vedic Saraswatī, originating in the Himālaya, flowed through the Ghagghar and was joined at Bhatner in Hanumāngarh district by the Vedic Satluj (Fig. 4.3a). The other tributary of the Himālayan-born river was the Proto-Yamunā which then flowed through the Chautang Nālā and met the Saraswatī near Sūratgarh. Arguing that the small Siwālik rivers could not have enough discharge to supply all waters of the Saraswatī, he identified its source in the Himālaya. Sometimes later, this Himālayan river was diverted and the Saraswatī became dry by 3300–3000 yr B.P., as testified by post-Harappan Early Iron Age settlements.

An agricultural engineer working all his life in southeastern Sindh (Pākistān) M. H. Panwar on the basis of extensive fieldwork and analysis of aerial photographers for irrigation purpose described lucidly in his article in 1986 (appeared in internet in 2009) the course of the Saraswatī through the Pākistān part of the Thār and related it to the palaeochannel features in the Indian part of the desert (A. Kar, personal communication, 2015).

Archaeologists' Deduction

Exploring the territories of Bikāner (Rājasthān) and adjoining Bahāwalpur (Pākistān) Stein (1942) noticed clusters of prehistoric settlements on the banks of a dry channel, the width of which over a stretch of 160 km was nowhere less than 3.32 km and at places as much as 6.44 km, and its bed made up of firm loamy soil. It was the channel of the Hākrā, the downstream continuation of the Ghagghar. He was impressed with the popular belief of the local people that in the past there were ferry crossings across the 5-km wide river at Mathula in Bikāner and at Minār Paltanmunara in Bahāwalpur where the river was still wider. Stein also noted that wide channel meeting the Hākrā at Walhar was "unmistakably representing the ancient winding bed of the Satlej". He regarded the Hākrā as the true Saraswatī extolled in the hymns of the *Rigved.*

Upstream of the Hākrā, on the banks of the 5–10 km wide Ghagghar, Ghosh (1952) discovered the ruins of a number of settlements, a few of which he recognized as belonging to the Harappan Civilization. He deduced that through this wide water course once flowed a big river formed by the mingling of many rivers "not only geographically, but also culturally". He averred that the river could *not* have been dead during the lifetime of the Harappan civilization.

The main tributary of the Ghagghar is the Chautāng, another wide, dry water course. Investigation by Bhān (1972) led to identification of a number of Harappan sites on the banks of the Chautāng including at Siswāl, Mitāthal, Rākhigarhi and others. Bhān recognized the Chautāng as the palaeochannel of the Drishadwati, the main tributary of the legendary Saraswatī.

In the years that followed, the floodplains of the Chautang, the Ghagghar, and other tributary rivers were comprehensively investigated by archaeologists including Lal (1979, 1997, 2003), Thapar (1975), Bisht (1978, 1982, 1984, 1991, 2000, 2013) and many others. In the Hākrā reach, Mughal (1974, 1984, 1995) recognized 1360 sites including those of Harappan Civilization. According to him the large number of settlements dating back to the period 6000 yr B.P. to 3500 yr B.P., flourished along the river only because it was flowing perennially. Then, around 4600 yr B.P.—the beginning of the Mature Harappan Period—the capture of the Chautang (by the Yamunā) caused grave distress to these settlements. However, at Derawar the settlements survived because the Satluj continued to flow into the Hākrā through another channel further downstream. Concurring with the surmise of Mughal, Kenoyer (1998) (Fig. 4.3b) stated that towards the end of the Harappan Civilization, the ancient Saraswatī had totally dried up, because the original tributaries were captured by two other mighty rivers "The gradual drying up of the Saraswatī is documented geologically and in sacred Vedic and Brahmanical literature of ancient India."

Nearly 2378 settlements of the Harappan Civilization have been identified in the land watered by the River Saraswatī. The average size of the settlements during the peak of growth (Mature Phase) was 13.5 ha (Possehl 1999), and the largest

Rākhigarhi on the Drishadwati (Chautang) was spread over 135 ha (Shinde 2008, 2014; in Subramanian and Krishnan 2014). It goes without saying that only a major perennial river such as the Saraswatī, could have met the demands of the teeming people that inhabited in their large settlements.

Recognition by Geologists

R.D. Oldham (1886) of the Geological Survey of India believed that "the lost river" of the Indian desert was none other than the Saraswatī, a part of the water of which was contributed by the Yamunā River. According to Oldham, the Yamunā of the Himālaya after leaving the hills divided into two parts, one part going down south as tributary of the Gangā and the other part contributing its discharge to the Ghagghar–Hākrā in the Vedic times. The Satluj was the other contributory river of the Saraswatī. Its turning westward to form the Beās and the diminished rainfall were responsible for the loss of the legendary river.

In the 1887 *Imperial Gazetteer* it is stated that some of the earliest Āryan settlements in India from almost the Vedic time were on the banks of the Saraswatī.

In his comprehensive and scholarly paper C.F. Oldham (1893) showed that the Vedic river Saraswatī between the rivers Yamunā in the east and the Satluj in the west flowed in the beds of the today's Sarsutī–Ghagghar streams (Fig. 4.4). "Although the river below the confluence (with the Ghagghar) is marked in our maps as Gaggar,

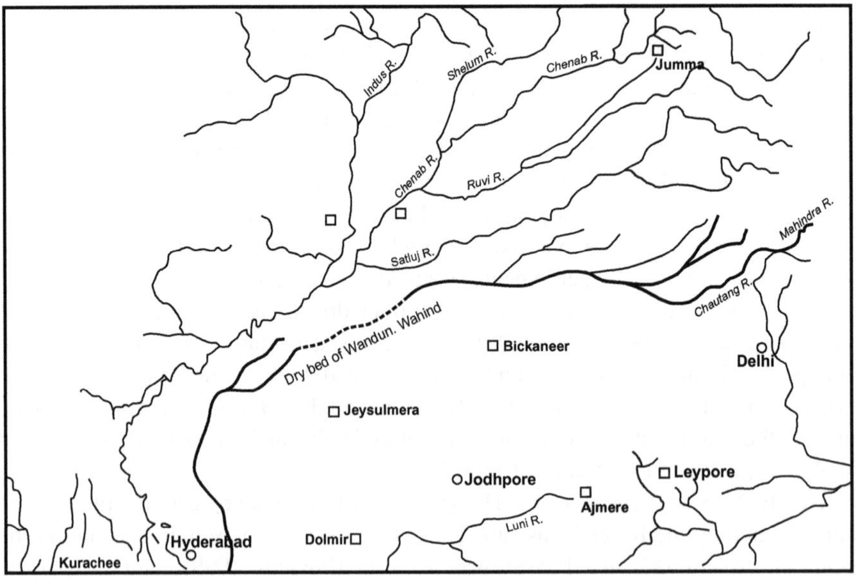

Fig. 4.4 Delineation of the Vedic Saraswatī by Oldham (1993)

it was formerly the Saraswatī. That name is still known amongst the people. It continued its journey across the Jaisalmer–Bahāwalpur region to the sea through the Hākrā and Nārā channels. The scattered mounds marking the sites of cities and towns throughout their tract imply that the country must have been fertile during that period". The main cause of the decline, according to C.F. Oldham, was shifting away of the Satluj—a river which has a history of changing courses. "The course of the lost river has now been traced from the Himālaya to the Rann of Kutch".

Sivewright (1907) regarded the Great Rann of Kachchh as the delta of the Hākrā River—"the lost river of Sind". The Gulf, according to him, was navigable until after the Arab conquest, and was later converted into the salt-impregnated marsh due to gradual silting up. The sediments that filled the Gulf must have been delivered by the Hākrā.

Raikes (1964, 1969) encountered at the depth of 11 m below the floodplain of the Ghagghar a "coarse greyish sand very similar in mineral content to that found in the bed of the present Yamunā"; and this sand body, traceable down to the depth of 30 m, extended over a width at least four times that of the bed of the present Yamunā. Significantly, above this sand horizon, greyish sand alternates with silty clay—the sediment containing "grey granite-derived material that occurs in the Yamunā". Implied is the postulation that the sediments were deposited by a Himālayan river.

Valdiya (1968, 1996, 2002, 2013) marshalled a wealth of information generated by the geomorphological, tectonic, sedimentological, hydrological, archaeological and remote sensing studies to reconstruct the life of the Saraswati formed by the confluence of the Shatadru (Satluj) and the Tamasa (Tons) and to sketch the landscape of the land it flowed through (Figs. 1.5 and 4.6). Nurturing a vibrant culture for over three thousand years, the Himālayan-born Saraswati discharged into the Gulf of Kachchh that then extended up to what is today the Great Rann of Kachchh.

In recent years a number of earth scientists including Ghose and Kar (1979), Courty (1995), Puri and Verma (1998), Kar (1999b), Thussu (1999), Saini et al. (2009), Kshetrimayum and Bajpai (2011), Sinha et al. (2012), Chaudhuri (2012), Van Dijk et al. (2016) and others through their comprehensive multi-faceted studies established beyond doubt the existence and the dimension of the River Saraswatī. All this is discussed in the following pages.

To recapitulate, the Saraswatī River rose in the Great Himālaya and beyond. Its eastern branch, the Tamasā (Tons) flowed southwest through one of the tributaries of the present-day Ghagghar in the foothill belt and met with the south-flowing western branch of the Shatadru (Satluj) at Shatrānā, about 25 km south of Patiālā. Then onwards the mighty river that it had become flowed in a 6–8-km wide channel. The Ghagghar is known as the Hākrā in its western middle reaches and as the Nārā in the lower reaches.

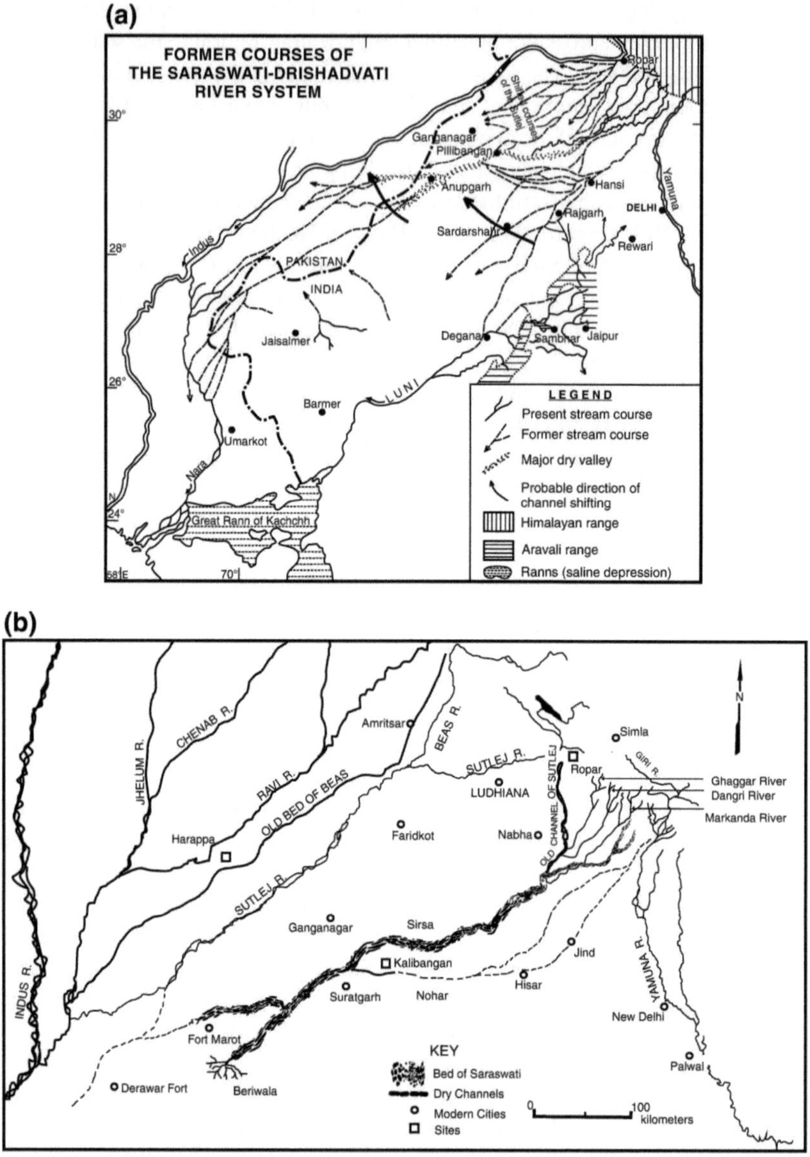

Fig. 4.5 Depiction of the Saraswatī by **a** Ghose et al. (1979) and **b** Yashpal et al. (1980) on the basis of analysis of satellite imagery

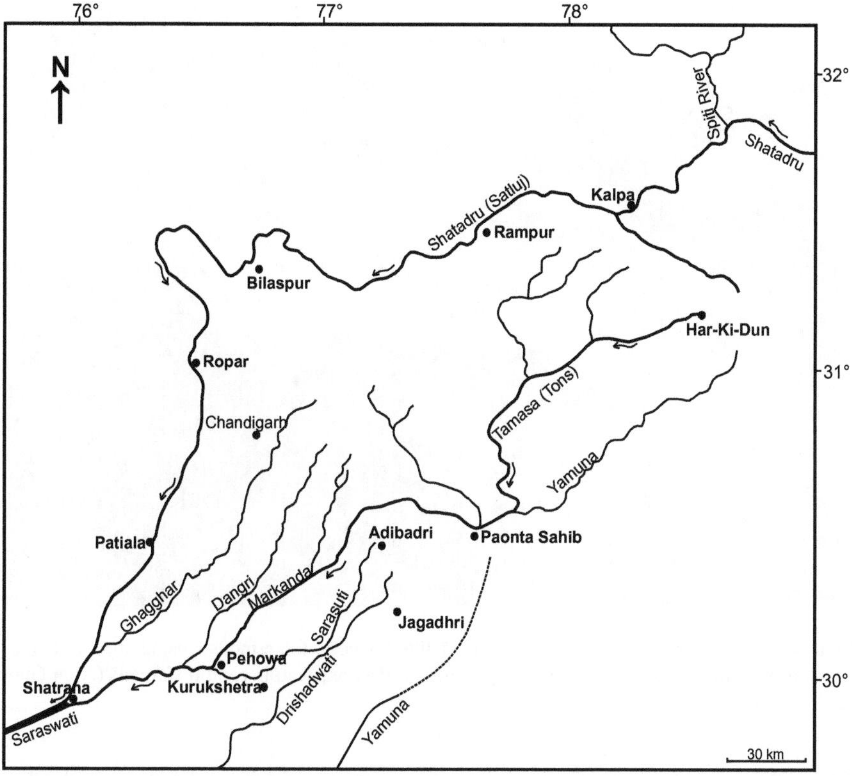

Fig. 4.6 The two branches of the prehistoric Saraswatī—the eastern branch was known as the Tamasā (now represented by the Tons) and the western was called Shatadru (now known as the Satluj)

Confirmation by Remote-Sensing Imagery

The bench-mark studies Ghose et al. (1979) and Yashpal et al. (1980) graphically brought out the network of channels of the Vedic Saraswatī as vividly revealed through interpretation of satellite imagery processed by advanced digital enhancing techniques (Fig. 4.5). Then followed a succession of works by specialists in remote sensing application (Sood and Sahai 1983; Kar and Ghose 1984; Bakliwal and Grover 1988; Ramasamy et al. 1991; Sahai 1999; Gupta et al. 2004, 2008, 2011 (Plate 4.1); Sharma et al. 2006; Bhadra et al. 2009; Sharma and Bhadra 2012) establishing beyond doubt the existence and the sweep of the River Saraswatī in the land that is today partly parched and partly a desert.

The satellite remote sensing data (Landsat ETM, IRS P3 MW,FS, RIS P6 MW FS, LISS-3 and LISS-4 as well as microwave Radarsat SAR) have proved very useful in recognizing the courses of rivers, active as well as defunct. Indian

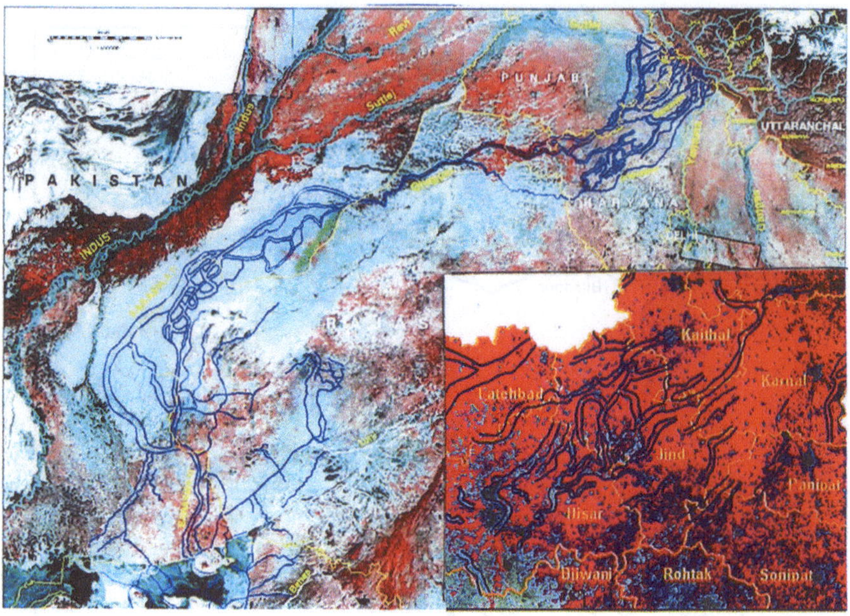

Plate 4.1 On the basis of interpretation of satellite imagery, the remote-sensing specialists mark the courses of the Saraswatī. *Inset* Palaeochannels of the Saraswatī in central Haryānā (After Gupta et al. 2008)

Space Research Organization carried out mapping of palaeochannels in the GIS platform using multi-resolutions and multi-temporal satellite images with piece-wise linear stretching techniques in parts of Haryānā, Panjāb, Rājasthān and Gujarāt (B.K. Bhadra, personal communication, 2015). The palaeochannels recognized and delineated on satellite imagery based map (Plate 4.1) exhibit remarkably close—almost distinct—connectivity with the present-day Ghagghar River in Hanumāngarh–Gangānagar districts (Gupta et al. 2004, 2008, 2011; Sharma et al. 2006; Bhadra et al. 2006, 2009, 2013; Sharma and Bhadra 2012).

It is evident that the river flowed in its multiple channels. The 75-km long palaeochannel between Ropar at the knee-bend of the Satluj to the point not far from the Ghagghar in Patiālā district is 1 to 6 km wide and dotted with archaeological sites (Fig. 7.1). Bhadra and Sharma (2012) describe the water courses as the palaeochannel of the Satluj that once flowed into the Ghagghar. Digital map processing of IRS P-6 AWiFS and Radarsat SAR images of the Great Rann of Kachchh clearly bring out a delta characterized by bird's foot pattern of drainage made up of intertwined channels (Sharma and Bhadra 2012). Significantly, one of these channels is traceable up to the Gulf of Kachchh to the south of the Rann. Obviously once the river that built the delta flowed straight south to the Gulf of

Kachchh. This delta is marked by areas of higher ground, standing up like island (*bets*) made up of fluvial sandy silt, and the depressions representing distributories and now filled with tidal clays (Malik et al. 1999).

Such ground information as multitudes of archaeological sites of settlements, drilling data, testimony of sedimentary bodies, hydrogeological data (groundwater level, water quality, undersurface discharge) and dating of water samples provide strong proof and thus validate the existence of these channels through which once flowed a perennial river (Interestingly, some minor palaeochannels in parts of Rājasthān that are rather away from the main Saraswatī palaeochannel along the Indo-Pākistān border, are bereft of Harappan settlements, because the rivulets were unrelated to the Saraswatī River).

Common Perception

Recognition by geographers, geologists, archaeologists and remote-sensing specialists apart, the common perception amongst the people of Haryānā is that the legendary Saraswatī flowed through their land. So etched is this understanding in the tradition of the people that the abnormally wide rivulets bear the names *Sarsutī, Saraswatī Nadī and Saraswatī Nālā*; and the road and railway bridges crossing the channels bear the name 'Saraswatī'. Some of these bridges were constructed and named more than a hundred years ago. In the government revenue records, the Saraswatī river is clearly shown as extending from Ādibadrī at the Siwālik foothills in Yamunānagar district to Pehowā in the Kurukshetra district. This reach of the practically dry channel is dotted with historical temples that draw thousands of pilgrims from far and near (Hitender Rao, *Hindustan Times*, May 16, 2015). In the Cholistan desert the people speak of the existence in the land of sand dunes of a great river that their forefathers crossed in ferry boats.

Danino (2010), in his very comprehensive, deeply absorbing and erudite work, summed up admirably what a large number of workers feel and believe "**If the vast majority of archaeologists and other scholars accept the Saraswatī's existence and location, it must be with good reason. Saraswatī was lost, but not forgotten**".

Chapter 5
Sediments Deposited by the River Saraswatī

Sources in the Himālaya

Emerging in the interior of the Himālaya and flowing through four lithotectonically contrasted and geomorphologically distinctive subprovinces or terranes of the northern orogenic (Himālayan) belt, as already pointed out (Fig. 3.1, Plate 2.1), the two branches of the Saraswatī with their tributaries brought (and continue to bring) enormous quantities of eroded materials and deposited them in multiple channels and in the floodplain. These Himālayan terranes have been rising up, but with variable rate, ever since the mountain had formed. The Great Himālaya is rising at the rate more than 7 mm/yr, the Lesser Himālaya between 3 and 5 mm/yr and the Siwālik a little more than 1 mm/yr (See Valdiya 2012 for references). The faster the uplift, the severe the erosion in the river valleys and the greater the volume of sediments brought and deposited by the rivers.

The fast-rising Great Himālaya, it may be recapitulated, is made up of Precambrian high-grade metamorphic rocks including muscovite–biotite–garnet schists, garnetiferous quartzites, kyanite-sillimanite gneisses and calc silicate gneisses (Figs. 3.1 and 3.2). These metamorphic rocks are intruded by Early Miocene anatectic granites. The eroded materials from the Great Himālaya therefore contain such resilient minerals as kyanite, garnet, biotite, muscovite and hornblende, besides a host of characteristic heavy minerals. The Lesser Himālayan mountain ranges are made up of Early to Later Proterozoic marine quartzites, phyllites, slates, dolomites, limestones and basic intrusives (dolerites) and volcanics (basalts) largely converted to amphibolites and chloritic schists and thrust over by piles of low-grade metamorphism and porphyritic gneisses. The Siwālik comprises Later Tertiary to Early Quaternary alluvial–fluvial sandstones, shales and claystones. North of the Great Himālaya, the Tethyan subprovince is made up of marine sedimentary rocks, ranging in age from Later Proterozoic to Eocene (Fig. 3.1). The northern part of this terrane is characterized by tectonically emplaced and squeezed

K.S. Valdiya, *Prehistoric River Saraswati, Western India*,
Society of Earth Scientists Series, DOI 10.1007/978-3-319-44224-2_5

up basic and ultra-basic rocks of mantle origin, intimately associated with deep oceanic sediments.

The Shatadru (Satluj) branch of the Saraswatī emerges south of Mount Kailās in the area of Mānsarovar–Rākshastāl lakes (Plates 2.2 and 2.4, Fig. 3.1). It originates not far from the great tectonic junction of India with Asia. It is the junction that marks the belt of docking of northward moving India with mainland Asia. The Shatadru has therefore very large catchment area and has thus been bringing sedimentary materials from Tibetan terrane as well as the four Himalayan lithotectonic terranes—the Tethys Himālaya, the Great Himālaya, the Lesser Himālaya and the Siwālik (Figs. 3.2 and 3.3). The Tamasā (Tons) branch of the Saraswatī, on the other hand drains only the later three terranes, for it originates within the Great Himālaya.

The Shatadru and the Tamasā together (Fig. 4.6) brought voluminous sediments on to the plains. The amount of detrital materials must have been considerably great in times when the Himālaya was convulsed in spasm of tectonism and rocked by great earthquakes. This is reflected in the nature and volume of sedimentary accumulation in their channels (fluvial deposits) and floodplains as alluvial sediments (Figs. 5.1, 5.2, 5.3, 5.4 and 5.5).

In recent years, a variety of geophysical investigations carried out and the drilling done for water and petroleum oil-gas have revealed the lateral extent and the thickness of the sedimentary deposits discernible in the plains of Haryānā and northwestern and western Rājasthān. These deposits must have been emplaced by a large Himālayan-born river (Valdiya 2002, 2013).

This is not to state that the Himālaya alone delivered sediments to the Haryānā–Rājasthān floodplains. The Arāvali orogenic belt also contributed weathered and eroded material. However, the proportion of the Arāvalian sediments could not have been much in view of the tectonic history of the hill ranges and the erosional history of this orogenic belt.

Sites and Modes of Deposition

The rivers deposit their loads of sediment as and when their carrying capacity declines. This happens when and where the velocity of the flow decreases. And the velocity decline may result from decrease in the gradient of the river-bed or reduction in water discharge. The decrease in river-bed gradient happens when downstream ground rises up or bulges up or due to faulting or folding, thus causing blockage. Dumping of landslide debris also brings about river ponding (blockage) and deposition of sediments. The sediments are deposited commonly along the

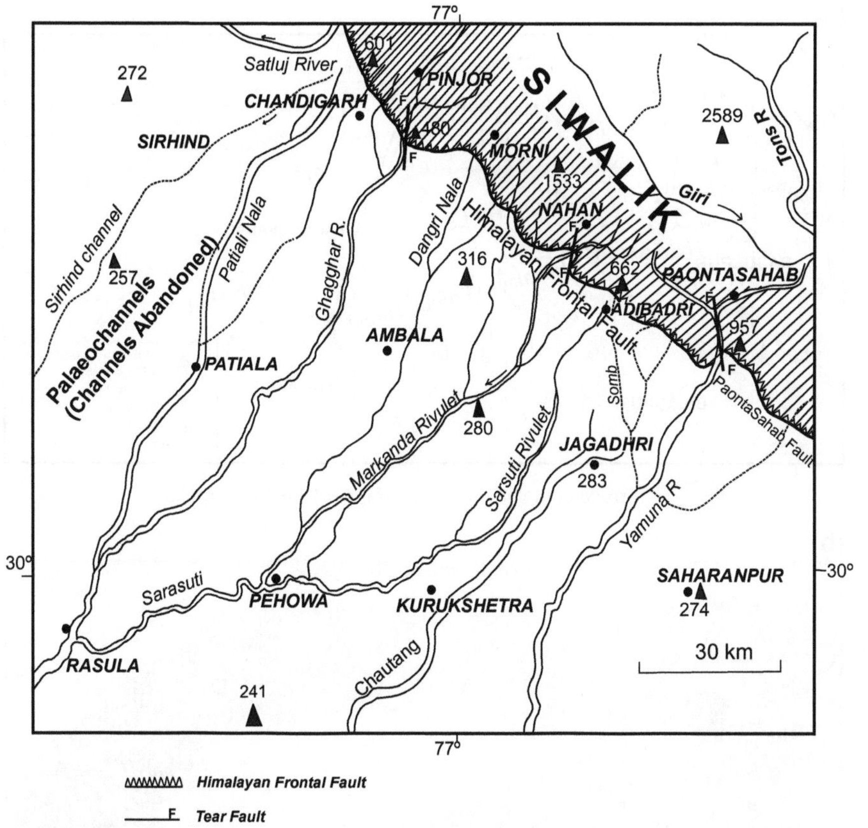

Fig. 5.1 In the northern reaches of the Saraswatī floodplain, the modern-day ephemeral streams represent the palaeochannels characterized by thick deposits of fluvial and alluvial sediments (From Valdiya 2002)

sides of the channels closer to the valley slopes which eventually form terraces, and at the bends of rivers as pointbar deposits. The terrace deposits contain remains of rocks, dominantly pebbles, cobbles and boulders transported by rivers from the source (provenance) upstream.

In the plains a part the sedimentary load, which is predominantly finer-grained sediments such as sand and silts, fill the channel, and another much finer part spreads on the adjoining plain during floods. The channel deposits have the geometry of half cylinders—straight or curving cylinders. The flood-deposits on the other hand form carpets or flat bands or tabloid bodies.

At their mouths the rivers dump fine sediments, mostly silts and clays, forming large fans called deltas. The deltas are characterized by numerous distributary channels that transport loads of sediments that rivers bring.

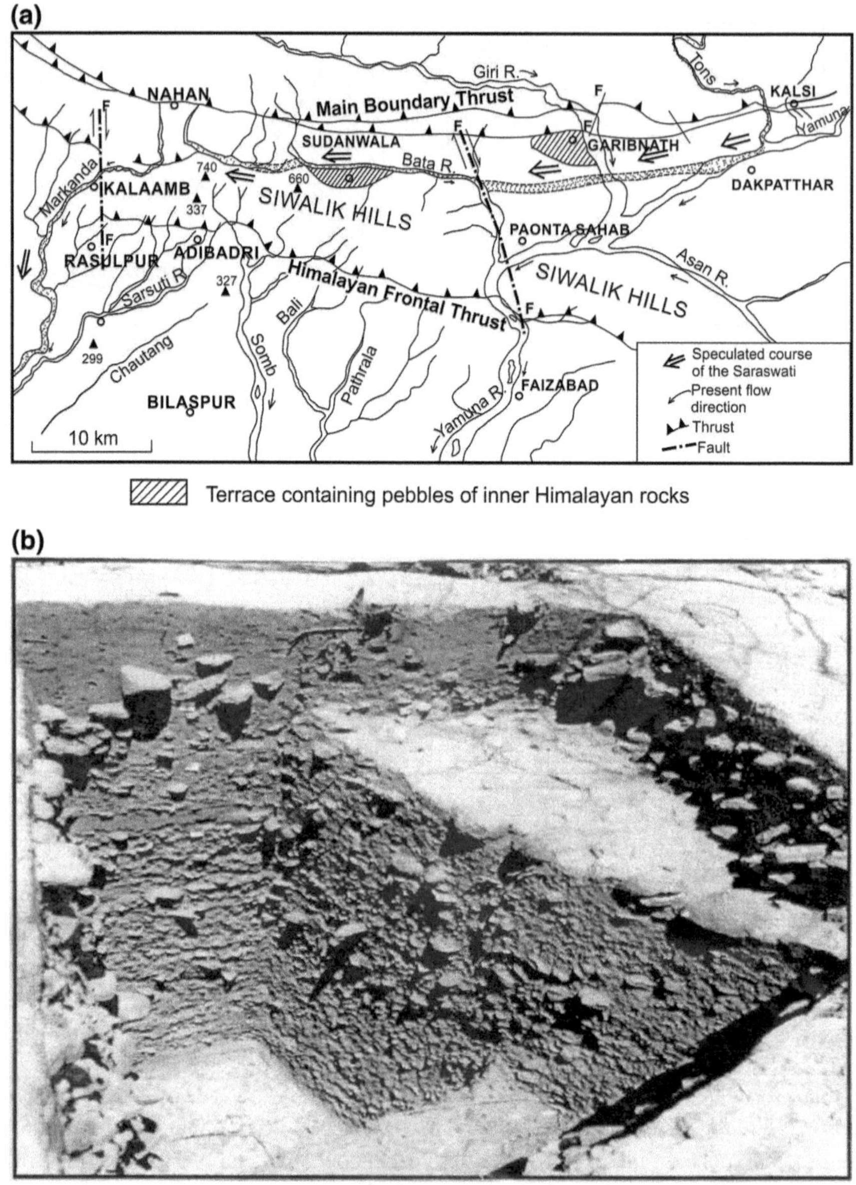

Fig. 5.2 **a** *Map* shows the fluvial terraces at Garībnāth and Sūdanwālā in the extraordinarily wide valley of the Bātā rivulet that originates and flows through the Siwālik subprovince. The *map* shows the present situation. The *arrows* show the direction the Tamasā followed prior to 3700 ± 250 yr B.P. (After Valdiya 2002; Puri and Verma 1998). **b** Clasts of metamorphic rocks of inner Himālayan parentage in the oldest (and uppermost) terrace of the Bātā—the valley through which once flowed the Tamasā

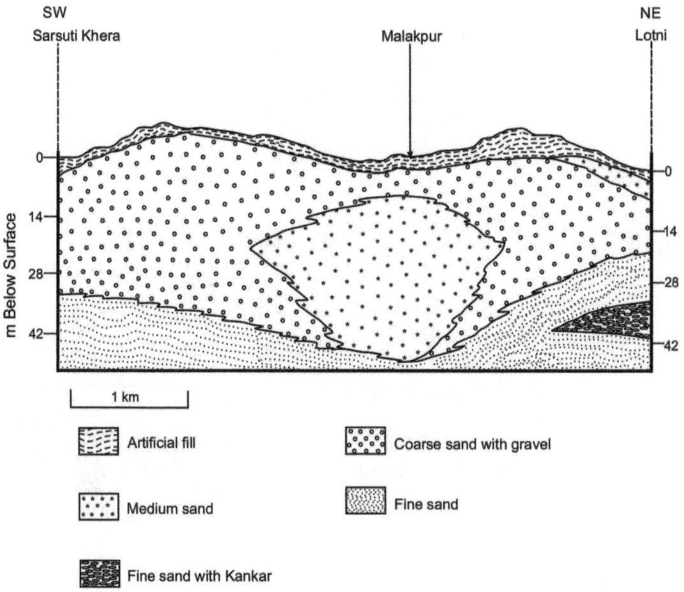

Fig. 5.3 Resistivity log-derived section between the Mārkandā rivulet and the Sarsutī rivulet. Both the thickness of the sand bodies and their lateral extent are considerable (After Kshetrimayum and Bajpai 2011)

Past Scenario

Siwālik Belt

Originating in and flowing through the Siwālik subprovince, the Bātā and the Mārkanda rivulets not only have anomalously wide valleys, but also are characterized by three levels of gravelly terraces, such as discernible at Garībnāth and Sūdanwālā (Fig. 5.2a). The terraces are 48–52 m above the bed of the Bātā. In the higher (older) terrace occur pebble-size clasts of schists, quartzites and phylites which have distinct affinity with the rocks of the inner Himālaya such as the Jutogh Formation and its equivalent in western Uttarākhand (Puri and Verma 1998). In sharp contrast, the two lower (younger) terraces are devoid of similar material but replete with clasts of the Siwālik—the formations that make up this subprovince or terrane. It is inconceivable that the Siwālik could have contributed metamorphic rock clasts to build the uppermost terraces of Garībnāth and Sūdanwālā. Only a river originating in the metamorphic terrane of the *inner Himālaya* could have brought and dumped them in the valley of the Bātā—once the course of the eastern branch of Saraswatī. However, more research has to be done to corroborate this inference. Detailed investigation of the terrace sediments, including the heavy minerals constituents, would demonstrate whether this inference is sustainable. More important is the precise dating of these terraces.

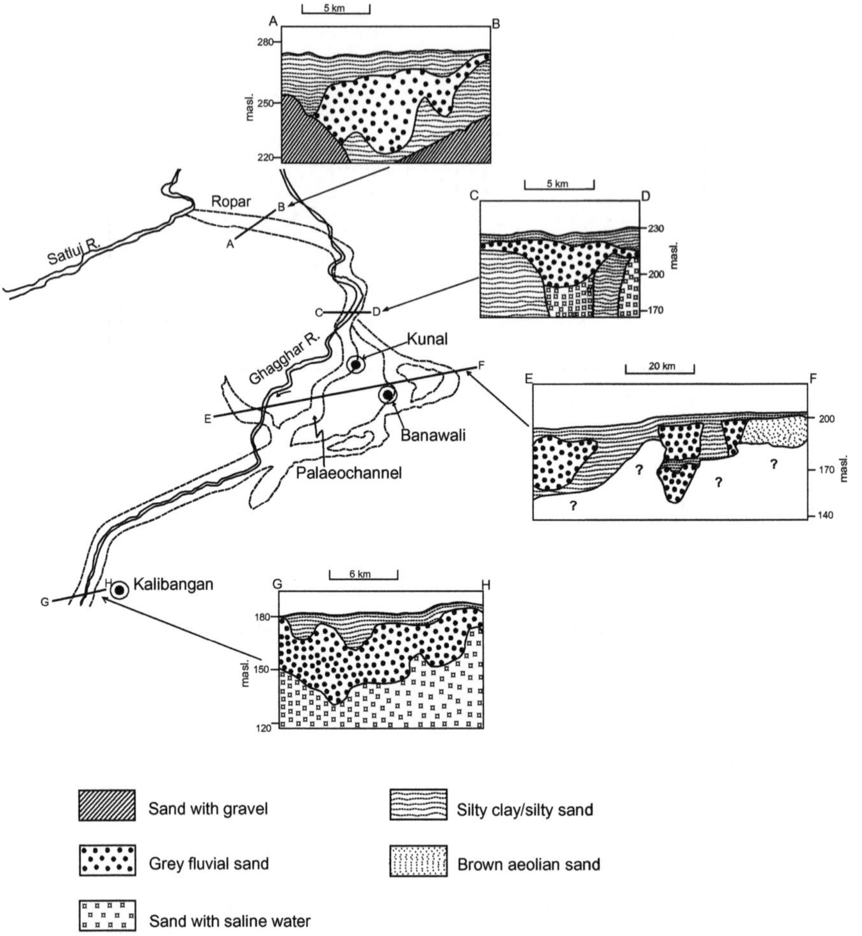

Fig. 5.4 The palaochannel of the Saraswatī River—including that of the course the Shatadru (Satluj) abandoned—is filled with thick and extensive sediments. Sections along the lines shown in the map (After Sinha et al. 2012)

In Foothill Plains, Northern Reaches

In the domain of the Mārkandā and Sarsutī rivulets (Figs. 4.6 and 5.1), vertical resistivity sounding applying Wenner configuration technique with maximum electrode spacing up to one km, together with the study of IRS ID LISS-IV (scale: 1:250,000) imagery revealed water-bearing buried sand bodies measuring 12 km in lateral extent (Kshetrimayum and Bajpai 2011). The sand bodies, composed of coarse sand and gravel at the depth between 48 and 148 m, connect the Mārkadā with the Sarsutī, implying intimate affinity of the two rivulets (Fig. 5.3). The lower course of the Mārkandā is hydraulically connected with the palaeochannels of the

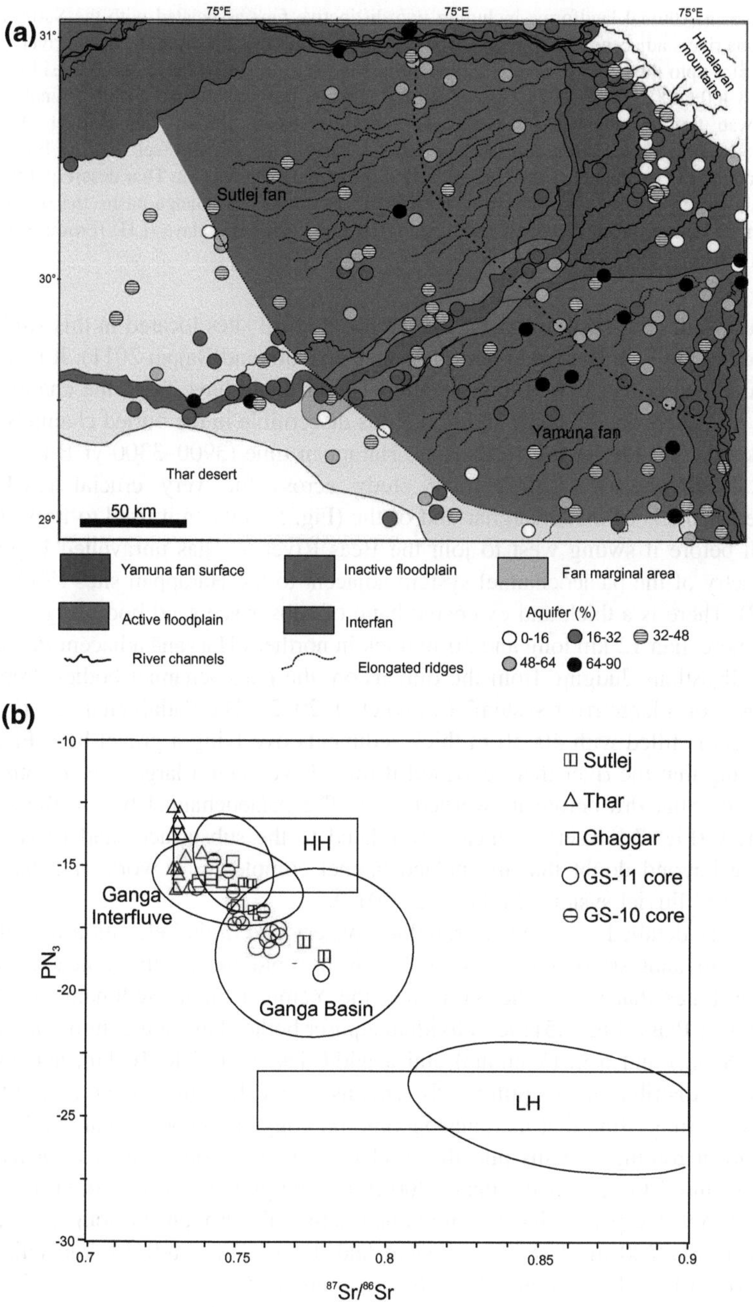
(a)
75°E
75°E
75°E
31°
30°
29°
Himalayan mountains
Sutlej fan
Yamuna fan
Thar desert
50 km
Yamuna fan surface
Inactive floodplain
Fan marginal area
Active floodplain
Interfan
River channels
Elongated ridges
Aquifer (%)
0-16
16-32
32-48
48-64
64-90
(b)
Sutlej
Thar
Ghaggar
GS-11 core
GS-10 core
HH
Ganga Interfluve
Ganga Basin
LH
PN_3
$^{87}Sr/^{86}Sr$
-10
-15
-20
-25
-30
0.7
0.75
0.8
0.85
0.9

◀ **Fig. 5.5** **a** Alluvial landforms including two distinctive fans associated with the Satluj and the Yamuna rivers adjacent to the Siwalik front. Now disconnected fans from the active river systems, these fans upto the depth of nearly 200 m formed in the long period dated 155,000±11,000 yr B. P. to 6,400±400 yr B.P. (Van Dijk et al., 2015) when the western and eastern branches of the Saraswati flowed into what is today the Ghagghar basin. [From: van Dijk et al., 2015]. **b** Neodymium versus strontium isotopic ratios clearly bring out the fact that the bulk silicate fractions of the sediments of the Ghagghar floodplain and channels, the Thar desert and the Satluj domain are no different from those of the Ganga interfluves and Ganga basin, the sediments of which are derived from the Great Himalaya (HH) and Lesser Himalaya (LH) [From: Singh, Ajit et al., 2016)

Sarsuti Nadi (Figs. 4.6 and 5.1). The archaeological sites located in this study area belong to the Late Harappan period (Kshetrimayum and Bajpai 2011). It is obvious from the study that only a large perennial river could have filled the channel with such thick deposits of sands and gravels as discernible in the buried channels of the Mārkandā and the Sarsutī in the Late Harappan time (3900–3300 yr B.P.).

A comprehensive geoelectrical study across the very crucial reaches of palaeochannels of the Ghagghar and of the (Fig. 5.1) when it used to flow straight south before it swung west to join the Beās River, — has unravelled large-scale geometry of the palaeochannel system adjacent to the Harappan sites (Sinha et al. 2012). There is a thick and extensive band of subsurface sand body (Figs. 4.6 and 5.4) more than 12 km long and 30 m thick in northern Haryānā adjacent Panjāb and NW Rājasthān. Judging from the dimension, the palaeochannel bodies "represent deposits of a large river system" (Sinha et al. 2012). The Patiālī channel (Figs. 2.2 and 5.1) is filled with 40–50 m thick sediments overlying a gravel bed (Fig. 5.4), implying that the river that deposited it must have been a large one. It must have been the Shatadru before it swerved west. The palaeochannel belt in the satellite imagery (Fig. 7.1 inset) is uncannily related to the subsurface sand body, establishing beyond doubt that the palaeochannel complex is a work of a large and long-lived fluvial system (Sinha et al. 2012).

A very detailed study, using satellite imagery and digital elevation model, coupled with analysis of data sets of 243 wells upto the depth of nearly 200 m, demonstrates that across the Satluj and the Yamuna fluvial sediment fans in the Ghagghar Plain (Fig. 5.5) the individual aquifer bodies have a medium thickness of 7 m (Satluj) and 6 m (Yamuna) and a width less than 5 to 10 km, and that the thickness distribution of aquifer body remains practically similar over different depth intervals, suggesting that the palaeogeomorphology and depositional conditions of "sediment routing systems into the foreland remained consistent over at least the time required to deposit the upper 200 m of stratigraphy" (van Dijk et al., 2015). Simply put, the palaeochannels remained active all through the long time period when nearly 200 m of sediments were laid down. This study by van Dijk et al. (2015) further shows that while the total number of aquifer bodies decreases downfan, the individual bodies do not become appreciably thinner, indicating that "the rivers on the fan systems maintained their water and sediment discharge over the lateral dimensions of the Satluj and Yamuna fans" upto about 250 km down south from the mountain front. It is also obvious that there was large-scale avulsion and abandonment of "channel corridors" (van Dijk et al. 2015). Implied is the deduction that the rivers that built the huge alluvial fans extending nearly 250 km down south from the Siwalik front flowed consistently for a prolonged period of time when nearly 200 m thick pile of sediments was laid down in the Ghagghar Plain.

Comprehensive geological study of sediments of the buried channel down to the depth of 40 to 50 m and dated ~75,000 years to the Recent in the Ghagghar Plain demonstrates that the fluvial sediments are characterized by strontium ratio ($^{87}Sr/^{86}Sr$) varying from 0.7365 to 0.7783 and neodymium ratio ($^{144}Nd/^{143}Nd$) between-14.6 and-19.07, suggesting that the sediments were derived both from the Higher Himalayan (Himadri) and the Lesser Himalayan catchment and "supporting the involvement of a river system originating in the Himalayan hinterland" (Singh, Ajit et al. 2016). *It is thus obvious that the sand bodies were emplaced in the Ghagghar Plain by a large river system originating in the Higher Himalaya (Himadri) through the palaeo-Satluj channel.* The temporal variation in the Sr and Nd ratios in the sediments suggests, according to Singh et al (2016), "shifting in the sediment regimes between the Higher Himalaya (low $^{87}Sr/^{86}Sr$ and high $^{144}Nd/^{143}Nd$) and Lesser Himalaya (high $^{87}Sr/^{86}Sr$ and low $^{144}Nd/^{143}Nd$) sources corresponding to changes in glacial cover as a result of major climatic shifts".

Kurukshetra Reaches

Significantly, in the Kurukshetra and Kaithal districts there are distinct palaeochannels of considerable dimension passing through Kurukshetra, Kaithal, and Sirsā as the satellite (IR AWIFS, LISS-III) imagery of 2004 reveal (Sharma et al. 2006; Bhadra et al. 2009). The sediment bodies mentioned below corroborate the findings of remote sensing specialists.

Thirteen kilometres west of Kurukshetra at Bhor Sayidān a palaeochannel, more than two km wide, exposes upper sequence of fluvial sand body, made up of 12 beds of chocolate brown sand, clay and mud, the upper bed containing fragments of Painted Grey Ware pottery dated 3375 yr B.P., along with brick-lined well (Chaudhuri et al. 2008b). In the vicinity of Kurukshetra, 20 bore-hole data define seven cycles of fluvial sediments exhibiting fining-upwards sequence over a range of 90–110 m of sands, silts and clays characterized by *kankar* (S.K. Lunkad, personal communication, 2010). The sediments were deposited by a meandering river that shifted its channel frequently.

Middle Reaches

Extensive geological field work combined with analysis of subsurface lithological data collected from exploratory bore-holes dug by the Central Ground Water Board and geophysical investigations of sediments in the middle reaches of Ghagghar between Tohānā and Ellenābād, demonstrate existence of multi-channel, multi-lateral system of a strong fluvial regime (Saini et al. 2009). The subsurface architecture of the palaeochannels and their floodplains in the Sirsā–Phaggu sector is 10–25 km wide. A palaeochannel oblique to the present-day Ghagghar has 9–29 m thick sediment fill (Saini et al. 2009).

In the Hisār–Sirsā tract the Quaternary sediments resting on granitic basement is 206–343 m thick, the clay deposit in the southern part being extraordinarily thick (Singh et al. 1988). In the much larger Fatehābād–Hisār–Sirsā tract the alluvial deposit varies in thickness from 200 to 400 m, the individual beds ranging in thickness from a metre to tens of metres (Fig. 5.5). Ribbon-shaped clay bodies laterally grade into grey to brown sand bodies. Significantly, underneath the silty clay deposits occur 5–15 m thick column of micaceous *sands of Himālayan origin* (Saini and Mujtaba 2011). These sandy units show regularity in the western part of the Ghaggar plain.

The OSL dating shows that the older sediments are 26,000 ± 2000 years to 21,000 ± 2000 yr B.P. old and the younger deposits occurring in limited parts are 5900 ± 300 to 2900 ± 200 yr B.P. (Saini et al. 2009). The younger sediments thus belong just prior to the Mature Harappan time (5000–4600 yr B.P.). It seems that it swerved to the north immediately thereafter to the location where there is the palaeochannel of the Saraswatī—on the banks of which are located the pre-Harappan Kunal, pre- and Mature Harappan Bhirana and Banāwali (R.S. Bisht, personal communication, 2015). Now far to the north of it flows the present Ghagghar.

A significant finding (Saini et al. 2009) is the presence of such heavy minerals as tourmaline, greenish brown amphibole, garnet, sillimanite, kyanite, ilmenite and biotite in brown sandy facies of fluvial sediments. The grey fine-grained micacious sands occupying different depths are "similar to modern-day sediments of mountain-fed rivers like the Yamunā and the Gangā".

There are lenticular deposits of Holocene lake sediments at a shallow level of 1–3 m (Bhatia and Singh 1998). These include lentoidof gypsum and limestone (Saini et al. 2009).

The channel-fill between Kaithal and Sirsā in the Ghagghar basin is more than 20 m thick. According to Courty (1995) it is characterized by deposits (Fig. 5.6) emplaced in six stages: (i) Accumulation of more than 10 m of *sediments derived from the Himālaya* by a river that had heavy discharge prior to 40,000 yr B.P; (ii) Decrease in discharge of the river as it *turned into a seasonal stream* prone to floods. There was reworking of the 6-m thick older sediments by dry winds during the period 40,000–20,000 yr B.P., (iii) Formation of 2-m thick extensive cover of wind-blown sand and silt in the period 20,000–12,500 yr B.P., (iv) Increase in the discharge of the Himālayan river as reflected in the deposition of typically *Himālayan-derived sediments*, (v) Progressive reduction in fluvial sedimentations and increase in wind activity, giving rise to loess deposition and formation of gypsum layers and concretionary carbonate deposit *kankar* in lakes and marshes that had developed 7000–5000 yr B.P. ago, and (vi) Formation of localized sand dunes and reduced seasonal flowing, leading to silting up of river channels in about 5000–2500 yr B.P. period (Courty 1995). In the Sirsā reach of the Ghagghar, the

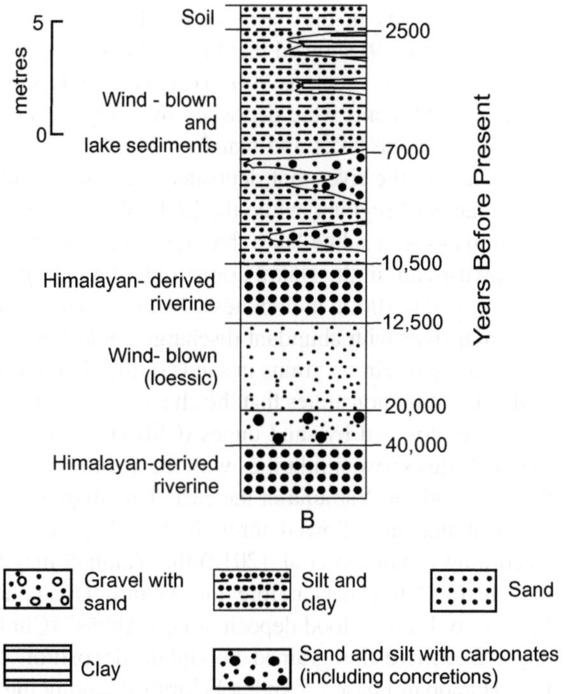

Fig. 5.6 Succession of sediments in the Ghagghar (Saraswatī) basin at a typical site. Note the presence of Himālayan-derived sediments in the Ghagghar tract (After Courty 1995)

grey sands containing constituents of Himālayan parentage are overlain by 7–8 m thick succession of loamy sands alternating with silty clays, the smectite clays having been derived from the provenance in the Siwālik. This means that in its later life of the river, the Siwālik was contributing dominantly to its sediment discharge.

Rājasthān Reach

In the Ghagghar floodplain in the region of Kālibangan, nearly 10 m below a horizon of mud and silt, there is more than 5 m thick horizon of carbonate-bearing micaceous sands. Dating of the shells and the OSL dating of sands clearly indicate deposition of the sands sometime during 9700 yr B.P and 5000 yr B.P. (Chatterjee and Ray, 2016). Moreover, the strontium ratio $^{87}Sr/^{86}Sr$ is more than 0.76, and the ϵNd(0) value minus 19 to minus 17, typical of the Higher Himalayan rocks (and definitely not of sub-Himalayan origin) (Chatterjee and Ray, 2016). These facts point to their provenance (source) in the Great Himālaya.

It is more than obvious from ample physical and chemical evidence that a major Himālayan-born river flowed along the Ghagghar channel—at least up to the Cholistān desert

Investigations in the basin north of the subsurface Barwāni–Jaisalmer Ridge (Figs. 3.5 and 3.9) showed that the thickness of the riverine sediments is more than 30 m—locally as much as 90 m (Raghav 1999). On the regional scale in the early stage, the sediments were deposited by a large river followed later by development of shallow saline lakes and marshes. Then there was dumping of sands by winds. Significantly, the river that deposited older sediments followed NNE–SSW trends of the faults of regions, while the ENE–WSW oriented faults mostly controlled the accumulations in saline lakes that developed quite later along them (Raghav 1999).

Downstream in the Hākrā domain, the fluvial sand body lying five metres below the river bed is 100 m in thickness, 14 km in width and 100 km in length. Only a big perennial river with abundant discharge could have emplaced this huge sand body. U-Pb dating of zircon grains in sands of the channels of the Hākrā (Ghagghar) north of the desert demonstrates that the channels were active until *after* 4500 yr B.P., and were later covered by sand dunes (Clift et al. 2012). Samples close to the archaeological sites show similarities with the sediments of both the Beās and the Satluj in the west and the Yamunā in the east. This finding corroborates the postulation that the Yamunā once flowed through the Ghagghar channel in the Saraswatī land. According to Giosan et al. (2012) the Yamunā may have contributed sediments to the Hākrā (Ghagghar) before the Mature Harappan phase, for they "recovered 5300 yr-old sandy flood deposit at Fort Abbās" (Cholistān) in Pākistān, and "on the upper interfluve fine-grained floodplain deposition continued until the end of the Late Harappan Phase". "And at Fakirābād among the dunes of the expanding desert are seen even younger sediments, approximately 3356 yr B.P. (Giosan et al. 2012). This is exactly what the author has been stating all along (Valdiya 2002, 2013).

Deposition at the Mouth

Through the channel of the Nārā the Saraswatī discharged into the sea now represented by the 350 km long and 150 km wide Great Rann. (Interestingly, the lower reaches of the Nārā around Amarkot is known as Hokdo to the local people who call themselves Hakdo). It was an arm of the Arabian Sea at the mouth of the Saraswatī—*the Rann Sea* (Plate 2.6, Fig. 5.7). A large delta complex represents the deposition of very fine material brought by the river. The delta spread far south to the Banni Plains, enveloping in its sweep the island of Pachcham. The Khadīr islet lies at the southeastern extremity of the delta (Fig. 5.8). There are two more deltas in the Rann Sea—one to the east formed by the Luni River, and the other to the west made by the Sindhu ((Fig. 5.7) which in the past emptied itself in the Rann sea). Study of clay minerals and geochemical analysis demonstrate that in the period 5500–2000 yr B.P., tidal energy was quite high in this sea (Tyagi et al. 2012). A "combination of climate and tectonic activity led to the withdrawal of the intertidal environment from the larger part of the Great Rann". It seems that the sediments deposited by the three rivers, accompanied by episodic tectonic uplift, transformed the embayment into a mud flat (Tyagi et al. 2012).

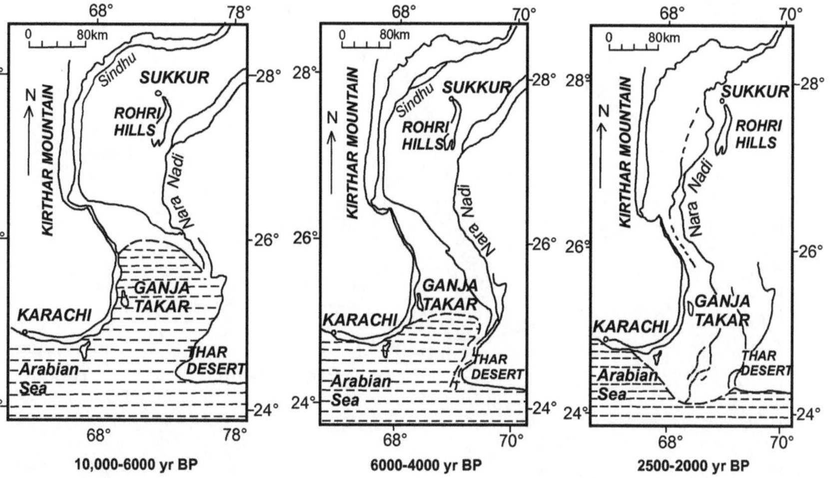

Fig. 5.7 The Nārā, representing the terminal reach of the Saraswatī, has been progressively building its delta 10,000–2000 yr B.P. ago and thus encroaching on the realm of the sea that is today the Rann of Kachchh (After Flam 1993)

Study of the Nārā channel indicates that water continued to flow through it after 2200 yr B.P; and in the interval 1200–1000 yr B.P., the sedimentation was dominated by both fluvial and marine processes (Ngangom et al. 2012).

The Saraswatī delta is characterized by a network of relict channels representing former distributaries (Fig. 5.8). These channels are now manifest as linear depressions filled with tidal clays impregnated with salt and covered by vegetation. The interfluves, on the other hand, are represented by raised ground or islets or *bet* of sandy sediments (Maurya et al. 2013). The delta sediments comprise very fine sands, silts and clays. Significantly, the heavy-mineral fraction of the delta deposit included rutite, zircon, hornblende, biotite and muscovite, presumably derived from the Himālayan provenance (Malik et al. 1999). It may be emphasized that a part of this delta is of Pleistocene antiquity and the Sindhu also contributed to its building in the past.

Faulting along an east–west fault in 1819 A.D. pushed up the northern part of the delta complex, giving rise to the 3–4 m high Allāhband Ridge (Fig. 5.8 inset) rendering the Rann sea unsuitable for navigation. Until then this sea was open to ships and boats. The palaeodelta complex unfolds a history of emergence and submergence in the past 4000 years, possibly due to neotectonic events during the Late Holocene (Merh 2011). Some workers attribute the emergence and submergence to sea-level changes (Gaur and Vora 1999; Gaur et al. 2013; Mathur 2002).

It may be emphasized that since the Rann belt has experienced strong and repeated tectonic upheaval, there is little chance of the original shape of the delta and its drainage pattern (of distributaries) remaining unchanged. There must have been considerable modification, if not obliteration. There is therefore strong need to revisit the area and study again intensively.

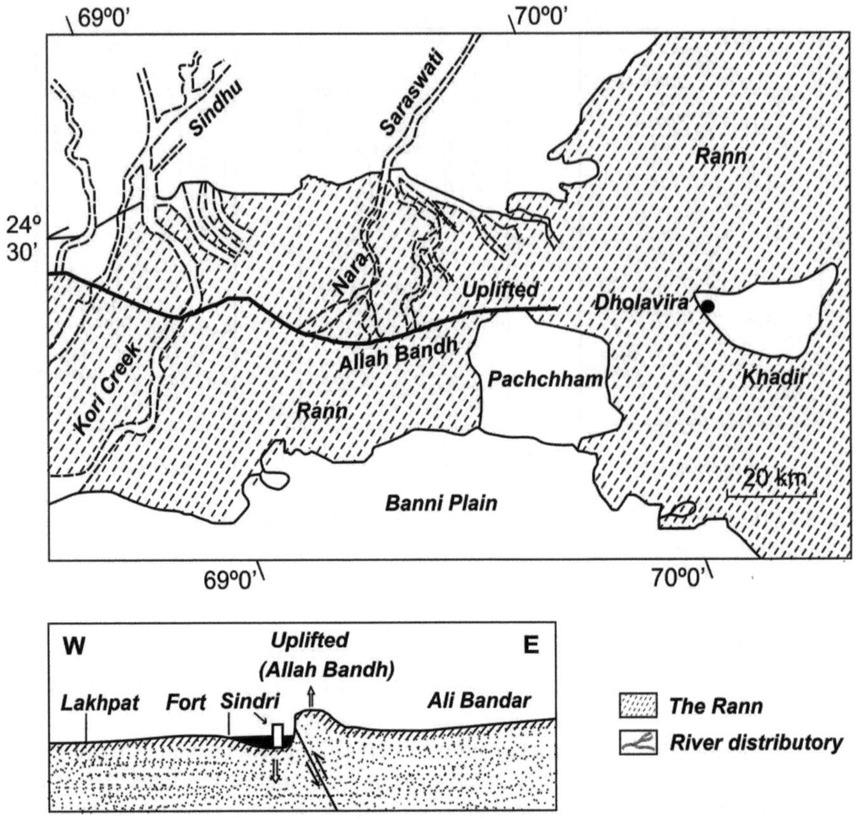

Fig. 5.8 The delta made up by the Saraswatī is characterized by a network of distributaries now filled with tidal clay deposits and covered by vegetation (After Malik et al. 1999). *Inset* Schematic section across the Allāhband showing faulting up of the ridge and sinking of Fort Sindri

Water Stored in Sediments

The heart of the Thār desert is ordinarily characterized by saline groundwater. Deep dug wells and tube wells have struck remarkably fresh (sweet, potable) waters from the aquifers of fluvial sediments, confined to winding-curving buried palaeochannels that crossed through the then rocky terrain of the Jaisalmer region. For more than 50 years the underground fresh water is being intensely used. But there was in 1998 no sign of the decline in the discharge and no indication of the lowering of the water table (Soni et al. 1999). Radar penetration of dry sands showed sensitivities of subsurface moisture which only a channel with active flow could provide (Rajawat et al. 1999). Electrical conductivity survey for ground water established the presence of palaeochannels (Fig. 8.2) replete with water (Kar 1999a). The relict channels are big, and seem to have served as the passage of huge discharges in this land of scanty rainfall. The channels thus recognized were presumably connected to the Ghagghar and Hākrā beds representing the ancient Saraswatī.

Electrical conductivity-based prospecting in the Jaisalmer–Hanumāngarh tract subparallel to the course of the Hakrā–Nārā has thus revealed the existence of fresh water at depths of 30–50 m and 60–250 m. Hydrogen-oxygen isotope chemistry of this groundwater shows a negligible content of tritium, the heavy isotope of hydrogen (Fig. 5.9) implying that this groundwater was not replenished through rainwater recharge (Rao and Kulkarni 1997; Nair et al. 1999). Significantly, in the Kishangarh–Tanot–Ghotaru tract the groundwater at the depth of 30–50 m was last recharged 1800–5000 years ago, and at the depth of 60–250 m in the time span 6000–22000 yr B.P., as radio-carbon dating shows (Nair et al. 1999). The 5000-year old water occurred in one of the palaeochannels delineated by ISRO (Bhadra, personal communication, 2015). Elsewhere, the age of this fossil at the depth of 50–60 m water is 1340–1850 yr B.P. (Rao and Kulkarni 1997). Interestingly, the subterranean water is flowing at an extremely slow rate—just 20 m in one year (Rao and Kulkarni 1997).

Across the border in the Cholistān the dry channel of the Hākrā River between Fort Abbās and Fort Mojgarh, fresh water has been tapped from a 100-m thick aquifer made up of slightly calcareous and micaceous fine-grained sand with subordinate mud (Fig. 5.10). Isotopically not different from the fossil water of the

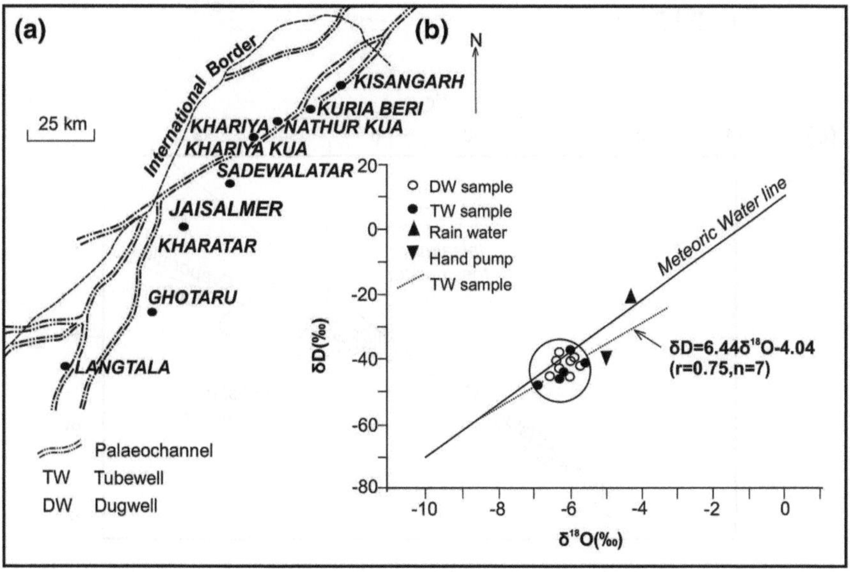

Fig. 5.9 **a** Palaeochannels delineated under cover of thick desert sands in the Jaisalmer region. These channels are remarkably close to the channel of the Hākrā across the border in Cholistān. **b** Hydrogen-oxygen isotope chemistry of underground fresh water (sweet) suggesting a distant source of water rather than a local rain water recharge (Nair et al. 1999)

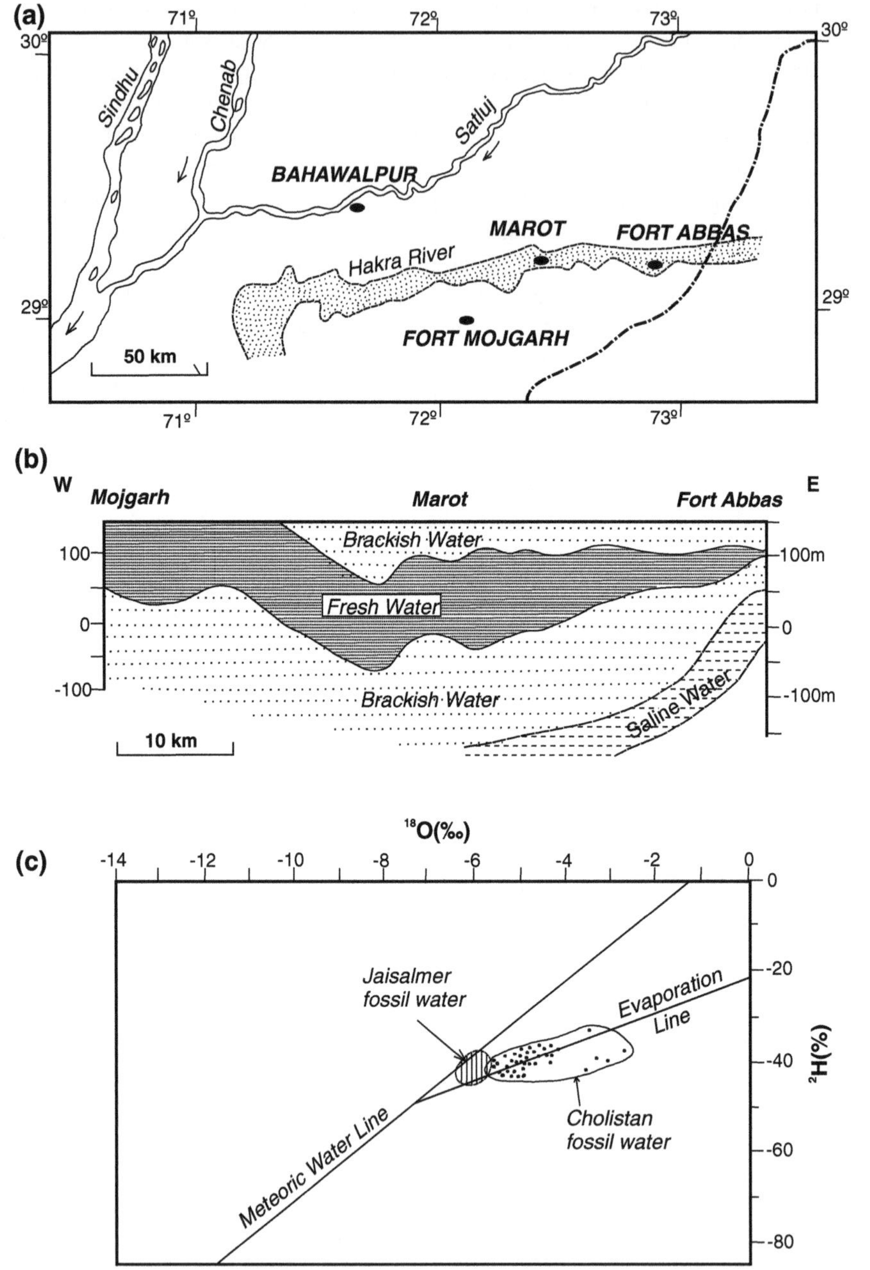

Fig. 5.10 **a** Hākrā floodplain in Cholistān (Pākistān). **b** Cross section shows the aquifer changes containing sweet water. **c** Plots representing Jaisalmer fossil water (Nair et al. 1999) inserted for the sake of comparison (Geyh and Plethner 1995)

Jaisalmer region, the Hākrā waters have yielded radiocarbon dates of 15900–7700 yr B.P., corrected to 12900–4700 yr B.P. (Geyh and Ploethner 1995).

The Long-Lived Saraswatī

Rivers Sindhu, Satluj, Karnāli, Arun and Brahmaputra originate in Tibet north of the great mountain barrier. They have cut deep gorges and canyons across the seemingly insurmountable range after range of mountains before descending on to the plain. These *antecedent* rivers (Fig. 4.6) are older than the mountain ranges they cross.

The thickness of the sediments throughout the Ghagghar basin from the Himālayan foothill to the Thār Desert, the testimony of the rounded clasts of inner Himālaya rocks occurring in the terraces within the valley of a Siwālik river, and the occurrence of very thick bodies of sediments characterized by heavy minerals of Himālayan origin in the sediments at various levels in the fluvial succession in the middle reaches and in the deltaic deposits at the mouth of the river Saraswatī, all point to the Saraswatī being a Himālayan-born river (Valdiya 2002, 2013).

Chapter 6
Peopling of the Saraswatī Land

Highlanders of Stone Age

The hilly and rocky terrains in the Saraswatī domain and the hill ranges of the Arāvali and the Himālaya that enclose it, are dotted with settlements of the people who made and used tools of stones and minerals. The makers of these implements were hunters and gatherers who had found water and raw material for making their stone artefacts in and around the Sarawatī land (Figs. 1.2 and 6.1). There is greater concentration of their settlements, understandably, in the rocky terrains of western Rājasthān and in the Siwālik hills drained by the rivers Suketī, Ghagghar and Mārkandā. While dependent on the harvest of forests, the Stone-Age aborigines seem to have lived in harmony with the nature as reflected in their paintings in caves such as those of Bhimbetka near Bhopal in Madhya Pradesh.

It is obvious from the shape and size of the stone implements (Fig. 6.2) that there was progressive improvement in the craft of making the artefacts through the long time span the archaeologists call Lower Palaeolithic (>700,000–100,000 yr B.P.), Middle Palaeolithic (100,000–32,000), Upper Palaeolithic (32,000–12,000 yr B.P.), Mesolithic (12,000–8500 yr B.P), and Neolithic (8500–7000 yr B.P.) periods. Needless to state, the Mesolithic and Neolithic periods belong to the Holocene Epoch.

Choice of Geological Material for Artefacts

The Stone-Age people chose their habitats where, apart from water and wildlife, raw materials in making tools were available. Rocks such as quartzites rhyolites and basalts were the preferred raw material. Amygdaloidal basalts contain such minerals as chalcedony, chert and jasper, very good for making tools. The Alwar Group of the Delhi Supergroup forming the bulk of the Arāvali ranges is made up of

K.S. Valdiya, *Prehistoric River Saraswati, Western India*,
Society of Earth Scientists Series, DOI 10.1007/978-3-319-44224-2_6

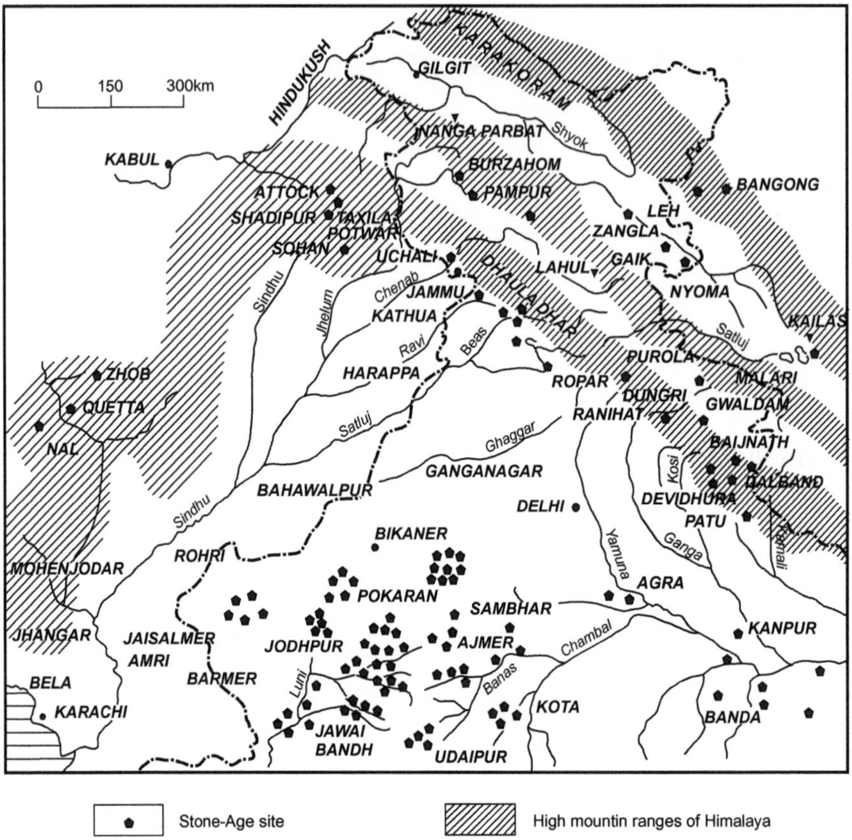

Fig. 6.1 Stone-age settlements in the Saraswatī and Sindhu basins. Note their concentration in the rocky terrains of western Rājasthāan and the hill ranges of the Arāvali and the Himālaya where raw material for making of tools was available (From Valdiya 2002; based on Misra 1961; Bose and Sen 1958)

quartzites and subordinate basalt and dolerite. Rhyolite is the dominant component of the Malāni Rhyolite occurring as big and small inliers in the vast expanse of western Rājasthān. Amygdaloidal basalt are the main constituents of the Deccan Volcanic Province that encompasses parts of Kachchh, Saurāshtra and Narmadā–Tāpi valleys in southern Gujarāt.

In the *Lower Palaeolithic* time they used rock quartzite and mineral quartz to make hand axes, cleavers/scrapers and choppers (Fig. 6.2). This period is also known as the *Acheulian Period.* The *Middle Palaeolithic* period witnessed making of smaller and lighter tools made of material struck out from rock rhyolite and mineral chert. Judging from the shape, size and diversity of these artefacts, it appears that there was a quantum jump in the progress of the people. For they seem to have perfected the craft of making implements for a variety of use. In the following period until the Neolithic time, the stone implements made of

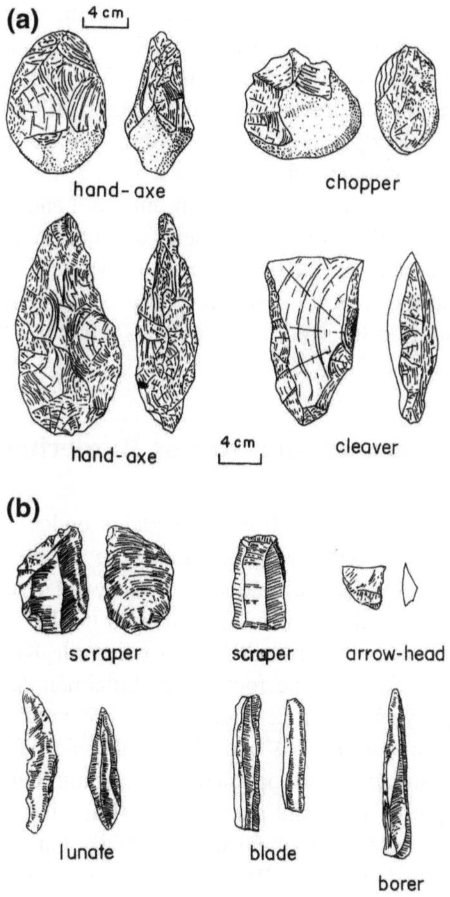

Fig. 6.2 **a** Stone implements of Acheulian (Lower Palaeolithic) period. **b** Artefacts of the Middle Palaeolithic, Upper Palaeolithic, Mesolithic and Neolithic times. (After Allchin and Allchin 1993)

chalcedony, jasper and agate were slender, sharper and faceted. Some had parallel faces, others were prismatic in shape and used as borers (Fig. 6.2). The forms and sizes of these *microliths* suggest that their makers were adapting to changes.

Time of Transition

It was the Neolithic period when the hunters-gatherers started domesticating animals and harvesting plants for food. Being in intimate contact for long with the nature, they had learnt the behaviours of animals and observed the phases of plant growth. And they understood the benefits of making use of them. The tool-making nomads began settling down in preferred locations and constructing abodes for dwelling. This was thus a time of transition—transition from the nomadic life to a settled life.

They settled down in places where the land was fertile enough for growing food plants, where water was available aplenty and where material for construction of dwellings occurred in abundance. The natural choice was the banks and floodplains of rivers and streams. The tool makers of hilly and rocky terrain thus descended on to the floodplains of the rivers of the Saraswatī system in the east and of the Sindhu system in the west.

A settled life in clusters meant more and closer interaction with neighbours and with people of other settlements around. Frequent exchanges generated social bond, culminating in the development of a social order. Eventually emerged a culture of preferred avocation and common belief. This was the beginning of a new civilization in the land of the rivers Sindhu and Saraswatī.

Immigrants from Across Bordering Mountains

The fertile and well-watered plains endowed with greenery and surfeit of food plants not only drew the Stone-Age people from the hilly and rocky terrains to floodplains of the Sindhu and Saraswatī rivers, but also attracted later on people from central Asian countries around the Pāmir massif described as *Meru Parvat* in the *Purāns* (Valdiya 2012). Crossing the Kirthar–Sulaimān ranges they came and settled down in the foothills of Sulaimān–Kirthar, the Makrān ranges and subsequently spread to the floodplains and beyond. Molecular genetic markers, including mitochondrial DNA and Y-chromosome data (Type 8–1), indicate affinity of the mainstream Indian population (including that of Sindhu–Saraswatī plains) with the people of west and central Asia (Barnabas and Suresh 1998) including northern Afghānistān. Turkmenistān and Tājikistān (Fig. 6.3). The immigrants came probably in the period 10,000–8000 years ago (Gadgil et al. 1998). There was thus induction of a new cultural element in the polity and culture of the original inhabitants of the lands of the Sindhu–Saraswatī, Sabarmatī and Narmadā rivers. The Sindhu–Saraswatī culture acquired dynamism and direction.

Stepping into New Age

Mehrgarh in the Kachī Plain and Kilā Ghul Mohammad in the Quetta valley (Balochistān) are believed to be the place where the village life first took roots (Fig. 6.3). Mehrgarh reveals evidence of continuous occupation from 7000 BCE to 2600 BCE (Mishra 2012). The Harappan civilization that is conspicuous by its absence there is strongly present just about 6 km away. Then the thread is picked up by Mehrgarh again where relics of early second millennium are found. At Mehrgarh on the bank of the Bolan River that descends from the Bolan Pass in the Kirthar Range, the people lived in rectangular houses built of bricks and mud. Some houses had a store room and open space. They cropped wheat and barley and reared cattle,

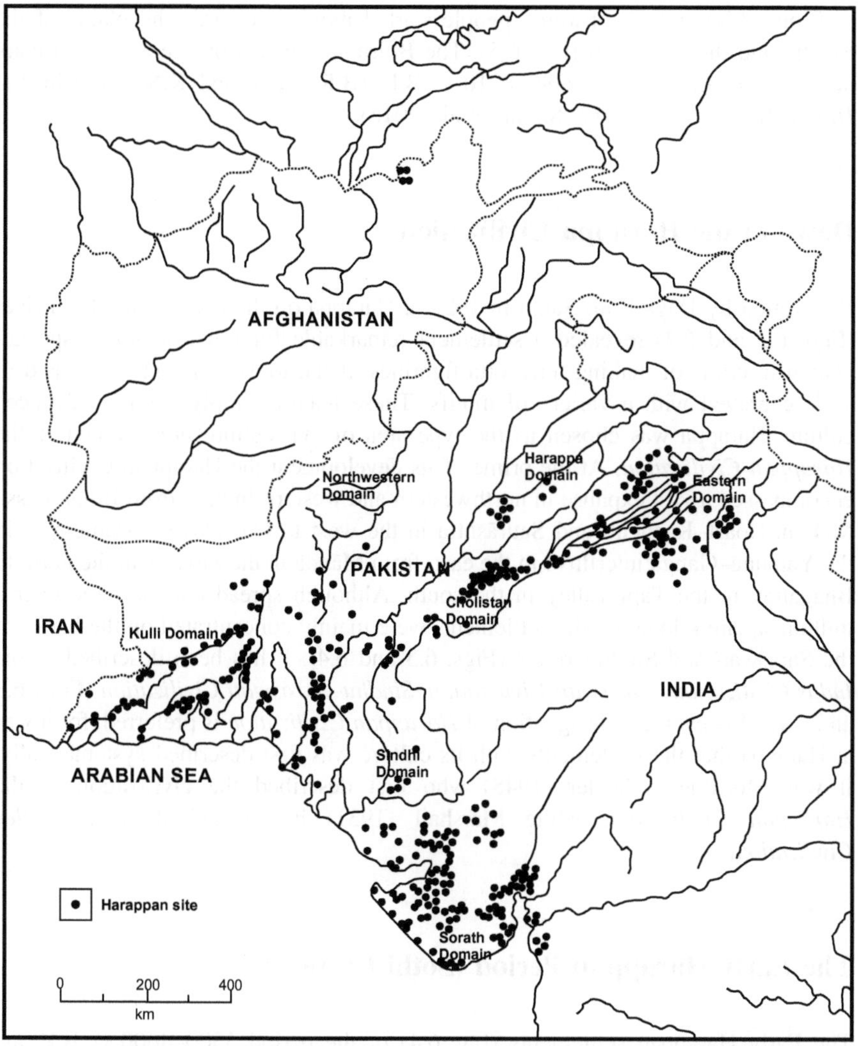

Fig. 6.3 Sites of Harappan settlements are more concentrated along the Ghagghar–Hākrā–Nārā water course, and also in the Kachchh and Saurashtra region (After Joshi and Bisht 1994)

sheep, goats and probably dogs and donkeys (Jarrige and Meadow 1980). The people buried their dead in flexed position, often with bodies of goat or sheep, and ornaments. Presence of such exotic materials as semi-precious stones as turquoise and lapis lazuli and of sea shells or locally available steatite indicates that they not only liked ornaments but also had trade relationship with people of faraway lands. They traded their cotton for semi-precious stones. This was the cultural scenario during the 9000–7500 yr B.P. That was time when there were no pottery and no metals. The Merhgarh settlements date back to the period 9000–5500 yr B.P.

From 6000 yr B.P. onwards people started moving towards the plains of the Sindhu and the Saraswatī (Fig. 6.3). The Bhirrana remains on the Saraswatī bank date back to the interval 9380–8210 yr B.P. (B.R. Mani and K.N. Dixit in *The Hindu*, November 4, 2012, Sarkar et al., 2016).

Dawn of the Harappa Civilization

Excavation by Daya Ram Sahni in 1922 at Harappa on the bank of the Rāvi river (Figs. 6.3 and 6.4) revealed a settlement remarkable for town-planning, specialization in crafts of making terracotta figurines, metal tools and ornaments, and of seals engraved with a variety of motifs. These features imply a fairly advanced culture. Harappa was chosen as the type area of that culture now known as the *Harappan Civilization*. At the prime of its development the Harappan Civilization encompassed a vast expanse in northwestern and western India—from Bolan Pass–Makrān Coast–Kachchh and Saurāshtra in the west to Chandayan–Ālamgīrpur in the Yamunā–Gangā interfluve in the east, from Māndā in the Siwālik in the north to Bhagatrav in the Tāpi valley in the south. Although spread over an area over a million square kilometre, the settlements were mainly concentrated on the banks of the Saraswatī and Sindhu rivers (Figs. 6.3 and 6.4). It has been described as the *Indus Civilization*, *Saraswatī Civilization*, *Sindhu–Saraswatī Civilization* (Figs. 6.3 and 6.4). However, the designation of *Harappan Civilization* is preferred, for it was at Harappa that the settlements with its culture was first described systematically. It was Mortimer Wheeler (1948) who first described the civilization as the *Harappan Civilization*, while Marshall (1931) had called it *Indus Valley Civilization*.

The Early Harappan Period (Sothi Culture)

The Early Harappan settlements (Fig. 6.4) of the period 5500–4600 yr B.P. are located at Bālākot, Amri and Kotdiji in the Sindhu domain, at Saraikholā and Gumlā in the intermontane Peshāwar plain, at Bhirrana, Ganweriwālā in the Hākrā plain in Cholistān, at Kālibangan, Kunāl and Banāwali in the Ghagghar domain, at Sothi Siswāl, Bhirrana, Mitathal, Farmānā and Rākhigarhi in the Drishadwati flood plain, and at Dholavīrā in the Khadir islet in Kachchh.

Spread over large area, Rākhigarhi was one of the five largest Harappan cities of that time (V. Shinde, in Subramaniam and Krishnan, *Frontline*, June 13, 2014). Comprehensively first surveyed by R.S. Bisht, Rākhigarhi has been declared at his initiative, a national monument of importance.

Among the assortment of features that characterize the Early Harappan Culture, the significant ones are the fortification of towns (Figs. 6.7 and 6.8), the brick walls, the pottery showing such motifs as intersecting circles, fish scale, and *peepal* leaf,

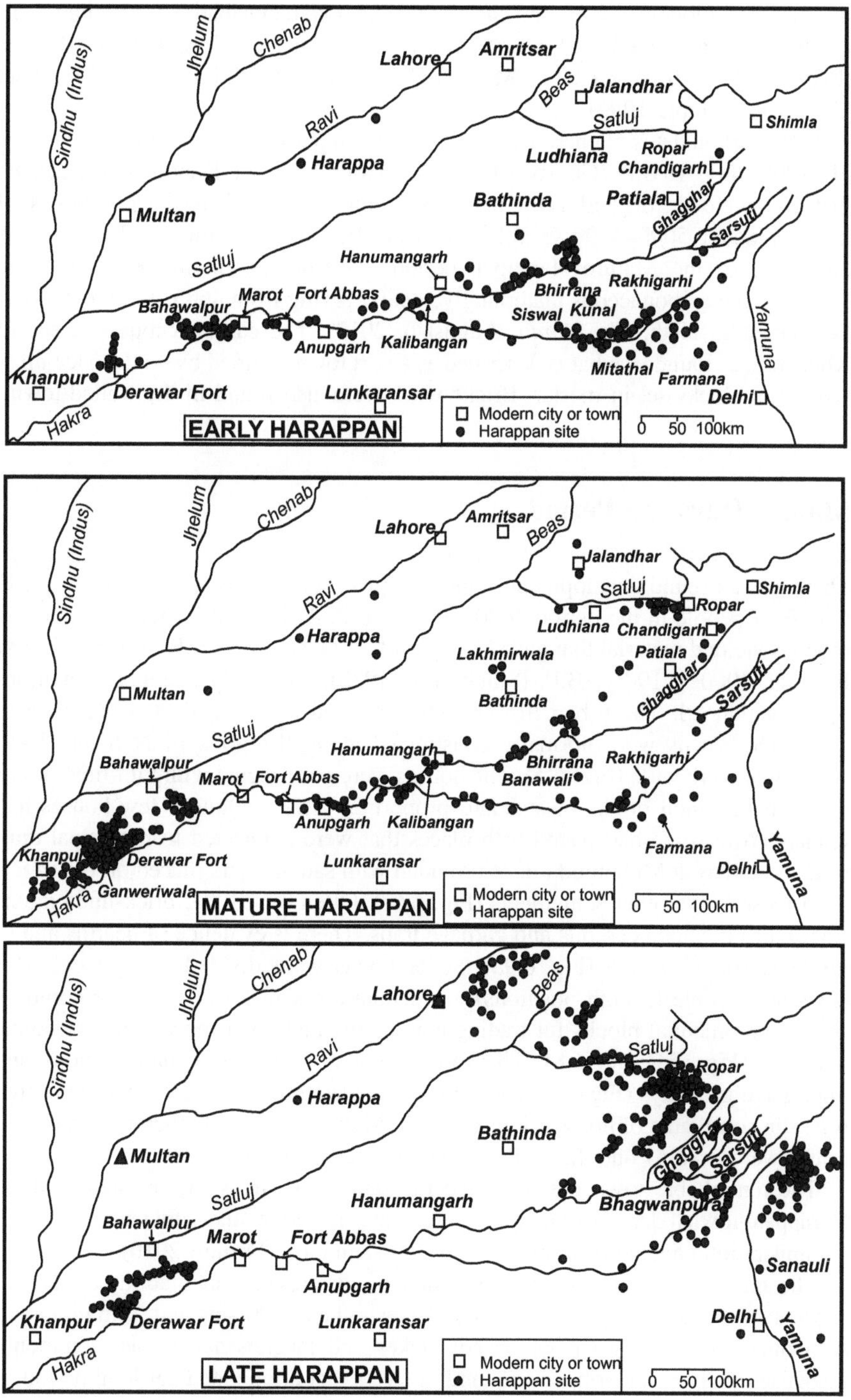

Fig. 6.4 Early Harappan, Mature Harappan and Late Harappan settlements in the period earlier than 5000 yr B.P. to about 3300 yr B.P. (From Danino 2010)

the terracotta models of carts and cartwheels, the ornaments made of carnelian, lapis lazuli, turquoise, shell and ivory, and the tools made of copper and bronze (Plate 6.1, Figs. 6.10 and 6.11) (Mughal 1974; Thapar 1975; Lal 1979, 1998; Bisht 1982, 1984; Gupta 1996).

Radiocarbon dates of charcoal from Kālibangan places this town in the time span of 5100 ± 800 yr B.P. (Courty 1995). It is Dholāvīrā in the Khadir islet, Kachchh that has yielded fortified settlements containing nearly 5-m thick deposits of material dated 5000–4500 yr B.P. (Bisht 1991, 2013)—a time well-before the beginning of the Mature Harappan period. At Khirsara on the Khari river in Kachchh, the carbonaceous material give dates varying from 4645 ± 45 to 4200 yr B.P. (Jitendra Nath, *The Hindu*, August 6, 2013). The carbonaceous material of Khirsara was found in what is described as a port town fortified by walls made up of sandstone blocks set in mortar. Bricks work were also found in the walled town.

Mature Harappa Period

There was all round development of the Harappan settlements and improvements in the life of people in the period 4600–3900 yr B.P. At Banwalli Ther the Harappan mounds yielded material that gave C_{14} ages of 3400 ± 130 yr B.P., 3900 ± 160 yr B.P. and 3900 ± 190 yr B.P. (Singh et al. 1988). The Early Harappan traditions were maintained. The urban life was characterized by sophisticated life style of elites. The towns were normally constructed along the grids of N–S and E–W streets that were wide (Fig. 6.5). The houses were constructed with kiln-fired bricks or sun-baked mud bricks of standard proportion (1:2.4). Quite a few houses had granary. And some had paved bath places that were connected to municipal drainage system as at Mohenjodaro, or provided with sanitary jars placed on the streets as in most of the other settlements (Fig. 6.6c). Often paved, the brick-lined drains (Fig. 6.6b) included cesspits and garbage traps. There is evidence of hearths and of *tandoors* (bread-ovens) (Fig. 6.9a). Some towns were divided into citadels (for elites and people in high positions), residential areas in lower town for common people, commercial blocks for trading community and open spaces for ceremonial purposes (Figs. 6.7 and 6.8). The fortification of towns now had bastions and impressive gateways (Figs. 6.7 and 6.8). There were dug wells lined with kiln-fired bricks in courtyards of houses (Fig. 6.9c). At Mohenjodaro and other sites trapezoid bricks were used to line the wells—some as deep as 15 m.

The remarkable town planning and the sanitary systems that existed in these Harappan towns demonstrate that there was an active and efficient system of administration capable of enforcing laws on sanitation (Danino 2010).

The amount of bricks used for construction indicates easy availability of clays in substantial amount. Clays occur in abundance in the floodplains and form important component of the deposits of ox-bow lakes and lakelets and pools, commonly associated with meandering rivers, and in rivers that are ponded for long period. It

Fig. 6.5 Houses the Harappans built. **a** Kālibangan (Early Harappan) (From Lal 2002). **b** Banāwali (Mature Harappan) (*Courtesy* R.S. Bisht)

has been shown elsewhere that both the Saraswatī flowing through the land in tectonic turmoil, was frequently ponded.

It is evident that the Mature Harappan period witnessed all-round progress in urban economy and in socio-cultural fields as represented by Mohenjodaro on the Sindhu, Harappa on the Rāvi, Ganweriwālā on the Hākrā (Saraswatī), Kālibangan and Banāwali on the Ghagghar (Saraswatī), Rākhigarh on the Drishadwati, Dholāvīrā in the islet of Khadir in the Great Rann of Kachchh close to the mouth of the Saraswatī and Lothal at the head of the Gulf of Khambhāt (Joshi 1972, 1990, 2008; Mughal 1974; Thapar 1975; Lal 1971, 1979, 1997, 1998; Bisht 1978, 1982, 1991, 2000, 2013; Rao 1991, 1999).

(a) (b) (c)

Fig. 6.6 Drainage system in Harappan towns. **a** Covered storm-water drain, Dholāvīrā (*Courtesy* R.S. Bisht). **b** Open drain made of bricks, Lothal (From Joshi 2008). **c** Cesspit container at terminus of a house drain, Kālibangan (Mature Harappan) (From Lal 2002)

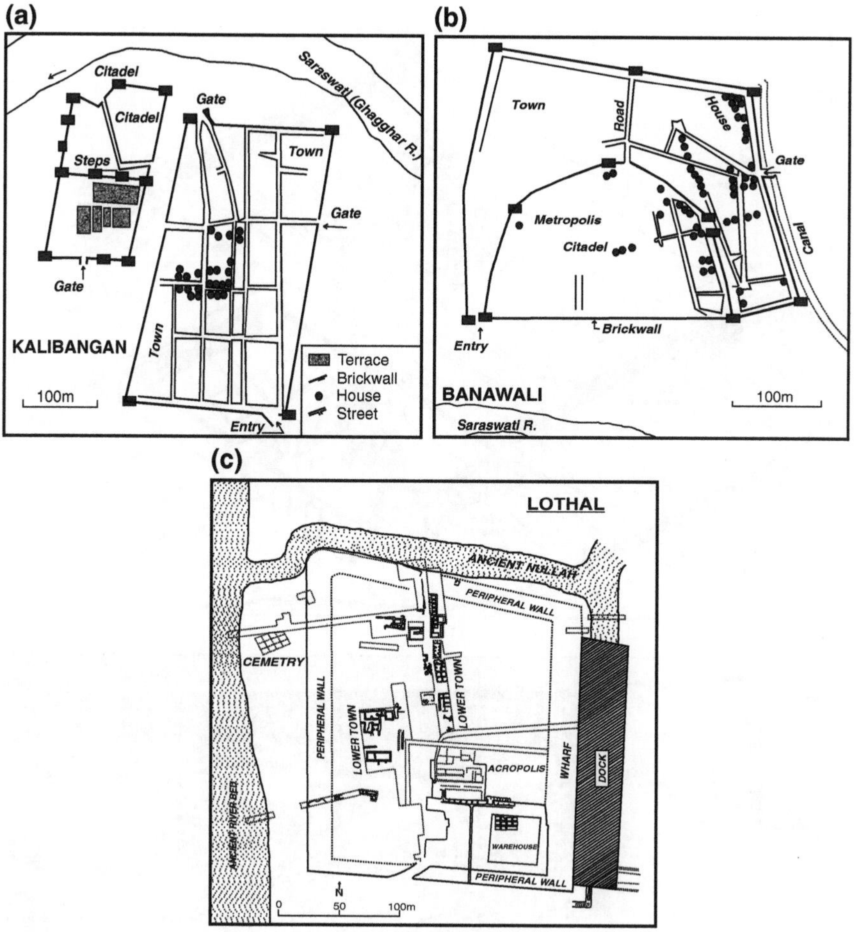

Fig. 6.7 Town planning in Harappan cities. **a** Mature Harappan Kālibangan (After Joshi 2008; Lal 1979, 2002). **b** Mature Harappan Banāwali (After Bisht 1984). **c** The port city of Lothal (After Khadkikar et al. 2004)

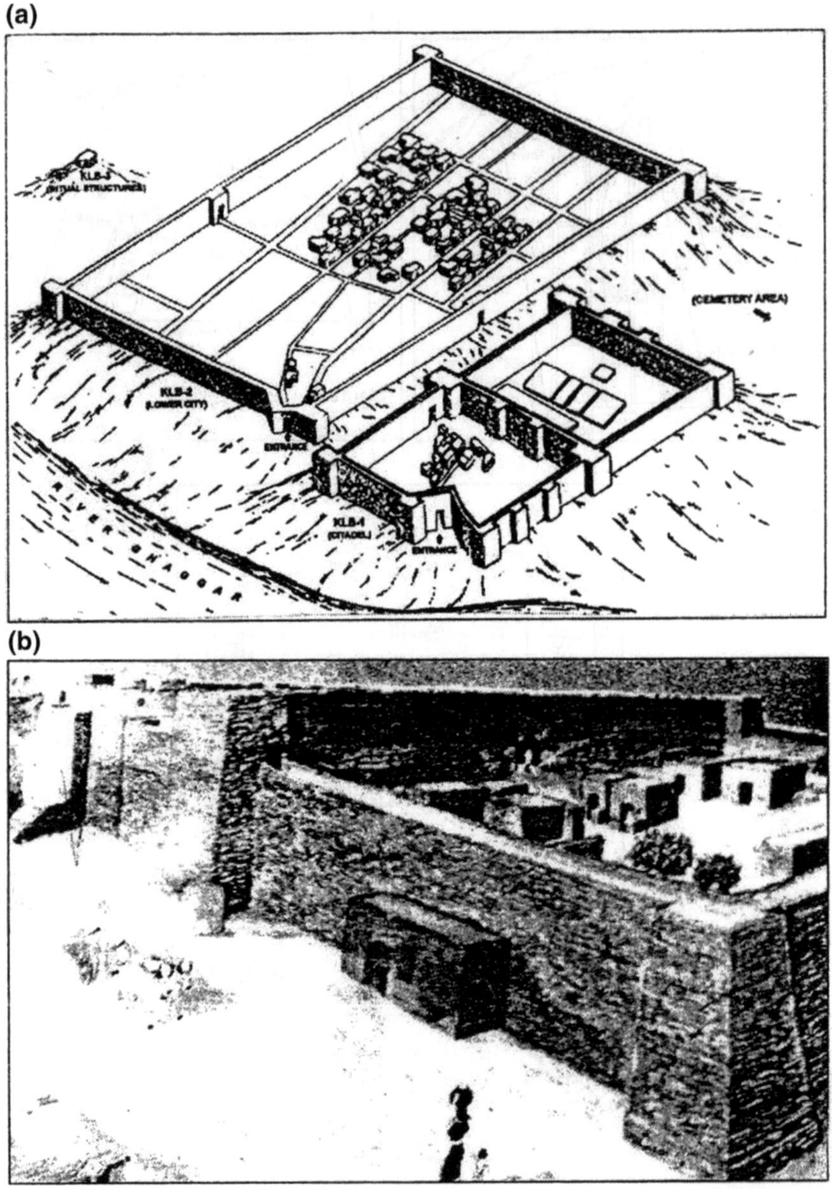

Fig. 6.8 **a** Kālibagan, the well-fortified town was divided into various sectors. **b** Well-fortified Surkotadā—an artist's reconstruction (After Joshi 2008)

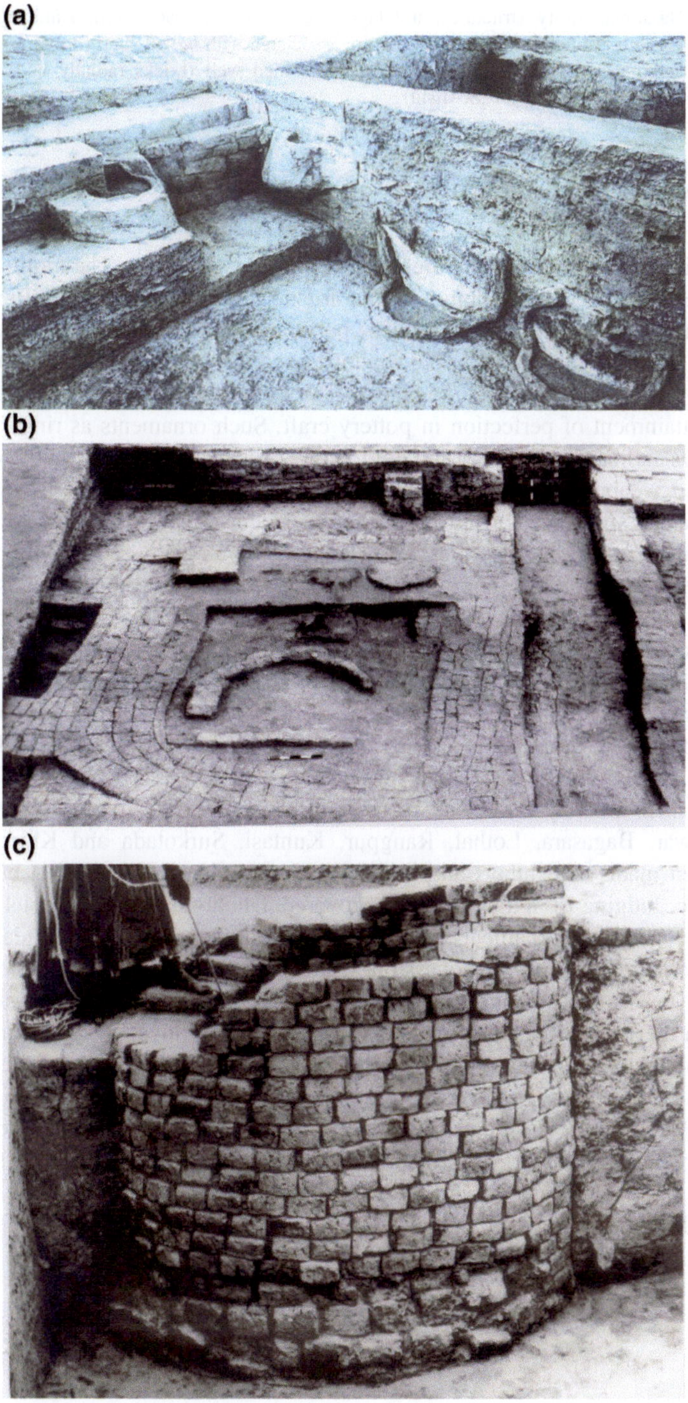
(a)
(b)
(c)

◀ **Fig. 6.9** Harappan utility structures. **a** Clay-made *tandoor* (bread oven), Kālibangan (Early Harappan) (From Lal 2002). **b** Apsidal structure associated with *havan-kund* (fire altar), Banāwali (Mature Harappan) (*Courtesy* R.S. Bisht). **c** Brick-lined well (bricks radially laid), Banāwali (Mature Harappan) (*Courtesy* R.S. Bisht)

Refinement in Handicraft

The making of pottery was the biggest and the oldest industry. The motifs (painted on pottery wares) speak volumes of the artistic taste of their makers (Fig. 6.10a). Their shape varied from bowls, goblets, perforated jars to beakers. Associated with 4230 ± 55 yr B.P. charcoal in the shallower sediments of the Lunkaransar lake in western Rājasthān occur ceramic pieces (Enzel et al. 1999), the ceramics testifying to the attainment of perfection in pottery craft. Such ornaments as rings, bangles, necklaces and beads of copper and gold, the vessels made of copper and bronze and tools and weapons of metals such as fish-hooks, chisels, axes, knives, daggers, arrow-heads, spear-heads, mirrors and bronze items (Plate 6.1) indicate not only their craftsmanship but also their knowledge of the metallurgy of converting ores into metals as borne out by crucibles and slags found with other remains. It goes without saying that they had the knowledge of mining techniques. For the people of the Saraswatī land, the nearest copper deposits must have been at Khetri in Rājasthān and the tin deposits at Tusham in Haryānā (Kochhar et al. 1999). The metallurgy of making alloy was known as early as 5000 yr B.P., as borne out by the sites of forging bronze and metal artefacts at Mohenjodaro, Harappā and Rangpur, Lothal, Dholavira and many other places.

The shell industry flourished in the coastal belt at such places as Dholāvīrā, Nageshwar, Bagasara, Lothal, Rangpur, Kuntasi, Surkotada and Khirsara. The craftsmen made shell inlays and bangles and also utensils like ladles, cups, jewellery etc. Judging from the remains discovered, Dholāvīrā was a vigorous centre of making beads, shell ornaments, copper wares and limestone pillars and slabs (Bisht 2013). Finding of semi-precious stones like carnelian, agate, jasper, bloodstone, quartz, steatite, vesuvianite-glossular, limestone, sandstone turquoise and lapis lazuli in the form of beads demonstrate the Harappans' love for ornament and for things aesthetic. Most important industry was the making of seals of stealite and terracotta which were engraved with motifs of animals, and some with a pictorial script (Plate 6.2b). The figures of animals in the seals demonstrate their concern or love for the animal world. Terracotta was made and used for making figurines, models, cakes, jars, bangles, and dices. Models of cart-frames, cart-wheels figurines (Fig. 6.11a), of a variety of animals and birds and horse at Surkotad Mohenjodaro,

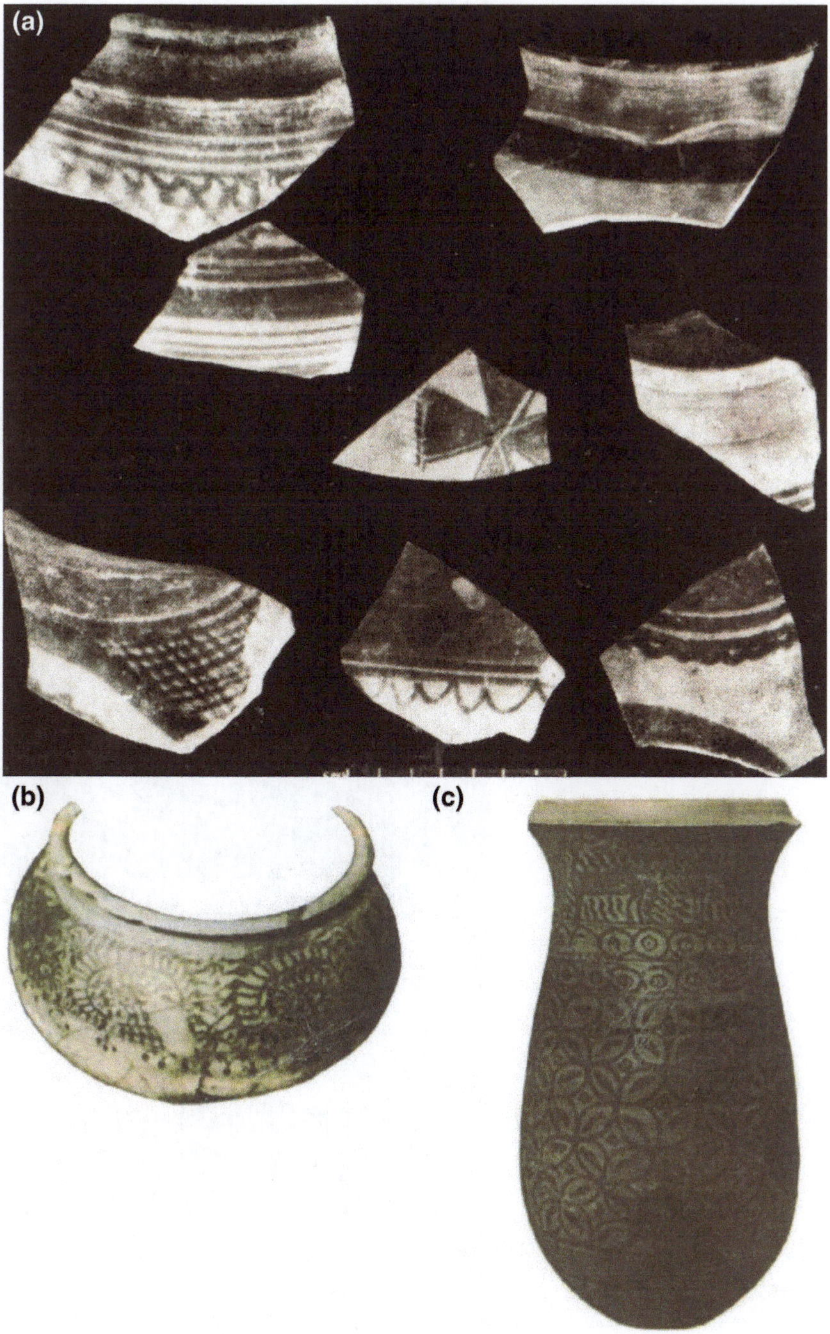

Fig. 6.10 Harappan pottery. **a** Pieces of painted pottery, Banāwali. **b** Painted pot, Banāwali. **c** Painted S-shaped jar, Banāwali (*Courtesy* R.S. Bisht)

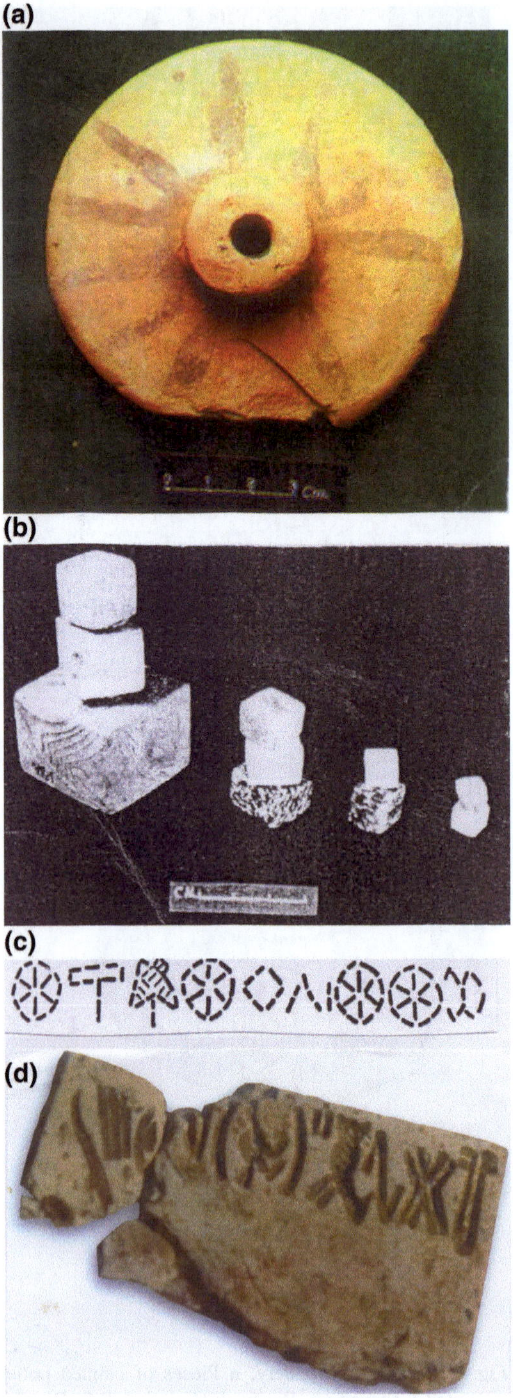

Fig. 6.11 **a** Terracotta model of cart wheel with radial spokes, Rākhigarhi (Mature Harappan) (From Lal 2002). **b** Scientifically designed measuring weights made up of mineral chert (After Joshi 2008). **c** Signboard on the gate of Dholāvīrā. Note the Harappan script (After Bisht 1998). **d** Harappan script on a tablet found at Rakhigarhi (From Subramanian and Krishnan 2014)

(a)

(b)

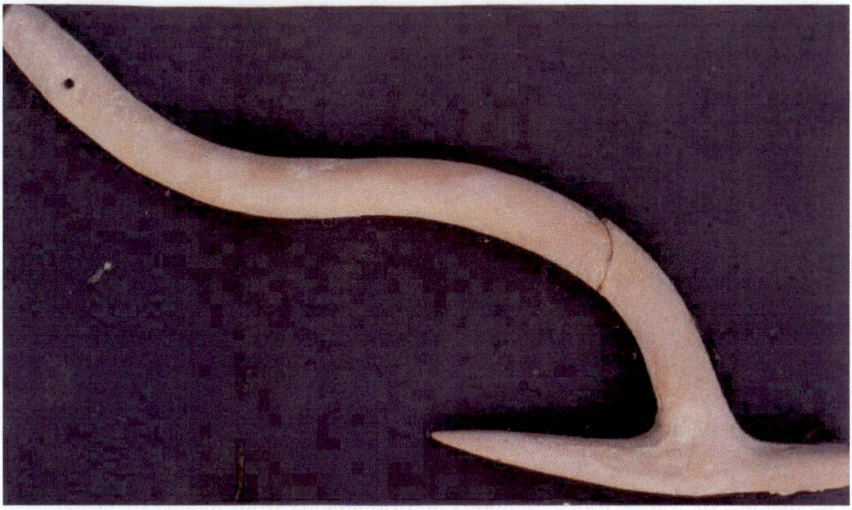

Fig. 6.12 **a** Furrow marks made by a plough outside the town of Kalibāngan (From Lal 2002). **b** Terracotta model of a plough at Banāwali (From Lal 2002)

Plate 6.1 Metal works of the Harappans. **a** Various copper and bronze weapons and tools, Banāwali (*Courtesy* R. S. Bisht). **b** Bronze bull, Kālibangan (Mature Harappan) (From Lal 2002). **c** Disc-shaped gold beads with holes along their diameter, Lothal (Mature Harappan) (From Lal 2002)

Lothal, etc. were made of terracotta. The model of cart-wheels indicate that the people used animal-driven mode of transportation.

Trade and Commerce

The making of wares, tools, weapons and ornaments from metals of beads, seals and figurines and models from terracotta, ornaments and utensils from sea shells, and of kiln-fired bricks from clays on large scale implies existence of demand-and-supply system. There must have been traders who procured goods from the makers/manufacturers and supplied to the users. Existence of weights for measuring mass provides a proof of trade and commerce in Harappan times. Made

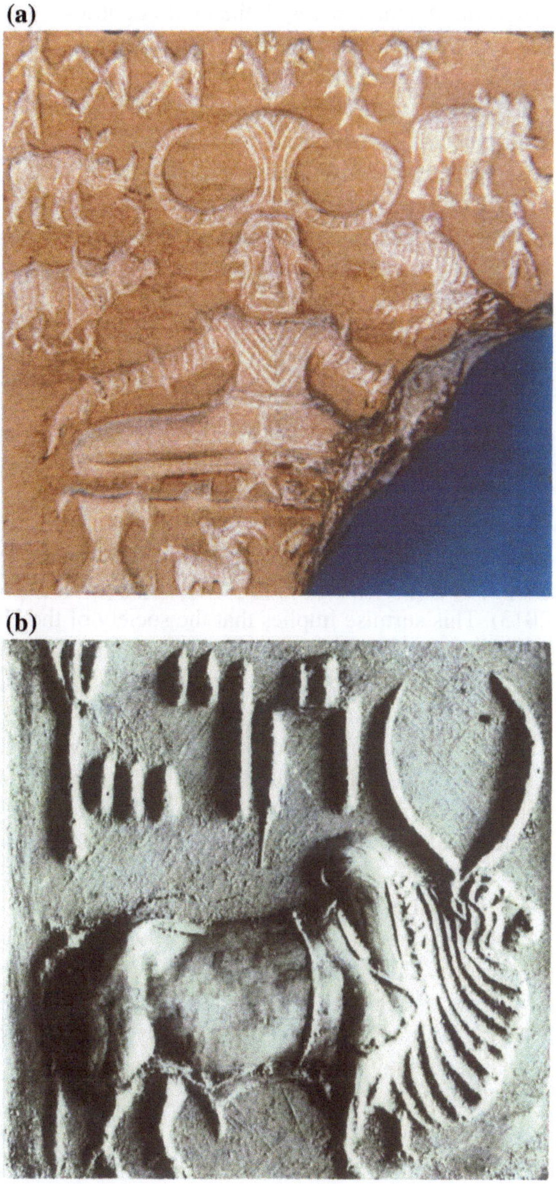

Plate 6.2 Harappan seals. **a** Motif of human figure in sitting posture surrounded by those of various animals (Mohenjodāro, Mature Harappan) (From Lal 2002). **b** Motif of humped bull (*Courtesy* R.S. Bisht)

of chalcedony, chert, blackstone (basalt and/or dolerite), the weights (Fig. 6.11b) were binary (1, 2, 4, 8, 16, …) in the lower scale and decimal in the higher.

The seals which bore inscriptions (Plate 6.2b) must have been used by traders for transaction of their business—inland as well as overseas. Dholāvīrā, Khirsara, Lothal, Padri, Bet Dwārkā were port cities inhabited with seafarers who traded across the Arabian Sea with the Gulf countries. This is borne out by the finding of seals of Gulf countries Oman, United Arab Emirates, Bahrein, Irāq and Elam (SE Iran) at Dholāvīrā, Bet Dwārkā and Lothal (Rao 1991, 1999), and of Harappan weights, pottery and beads bearing inscription in Oman and elsewhere. Such semi-precious stones as turquois, and lapis lazuli must have come from abroad, brought presumably by traders.

A pottery piece showing a boat found at Mohenjodaro indicates that there must have been inland navigation as well, a convenient mode of travel and transportation. It is not without significance that Rākhigarhi, Banāwali, Kālibangan, Ganeriwālā, among others, were all connected to the port city Dholāvīrā by the River Saraswatī.

A feature of tremendous import—rather of critical value—is the existence of script (Fig. 6.11c and Plate 6.2b) commonly used for business transactions and long-distance trade, as borne out by its uniformity across the Sindhu–Saraswatī domain. Seals bearing this script have been found at most of the places such as Mohenjodāro, Harappā, Dholāvīrā, Banāwali, Lothal, Kālibangan, Bhirrānā, Basor, Binjor, Khoisārā, Bagsara, Shikārpur, Surkotada and other places. Finds of Harappan relics at many sites in Mesopotamia, Susa and Bahrein testify to overseas trade connections. Decades of sustained efforts demonstrated that the Harappan script expresses the most popular language of the common man (Kalyanaraman 2013). This surmise implies that the society of the Harappan Civilization was quite literate. Amongst the most significant features found in Dholāvīrā is a signboard with inscription of 10 Indus signs, each a composite one (Bisht 2013). The signboard was presumably fixed on the facade of a gate (Fig. 6.11c, d).

Some cities are noted for special features. Mohenjodaro on the River Sindhu had a bathing pool, and its houses were built on raised brick-platforms as a measure to cope with flood hazards. Dholāvīrā was a port city in an isolated island (*bet*) Khadir in the Great Rann of Kachchh. All around the built-up divisions at Dholāvīrā, there were cascading series of reservoirs or tanks for storing water (Plate 6.4) conducted there from the monsoon channels of the Manhar stream on the south and the Mamhor stream on the north (Bisht 1991, 1997, 2013). Another port city Lothal at the head of Gulf of Khambhāt, there was a 4250 yr old dockyard (Plates 6.3 and 6.5) large enough to harbour many big boats and small ships (Rao 1973, 1985, 1991). The size of the Lothal dockyard was larger than that of the modern Vishākhapatnam (Mathur 2002). Nearby was a huge warehouse raised on mud-brick brick blocks and partitioned into halls, for stocking import-export cargos. People at Padri on the western coast of the Gulf of Khambhāt used to manufacture salt from sea water (Shinde 1998).

(a)

(b)

Plate 6.3 Port city Lothal. **a** Jetty filled with sand. **b** A warehouse with container for storing grains, close to the jetty

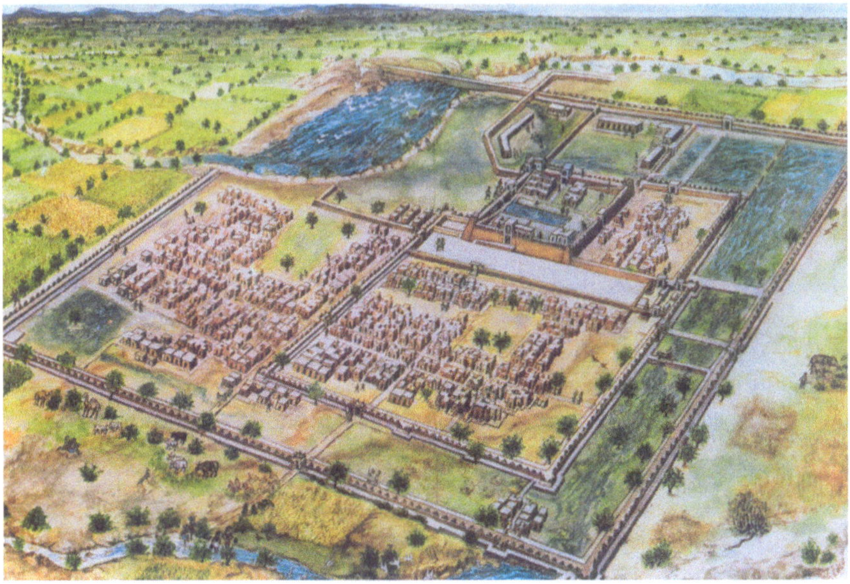

Plate 6.4 R.S. Bisht's concept of the vibrant port city Dholāvīrā

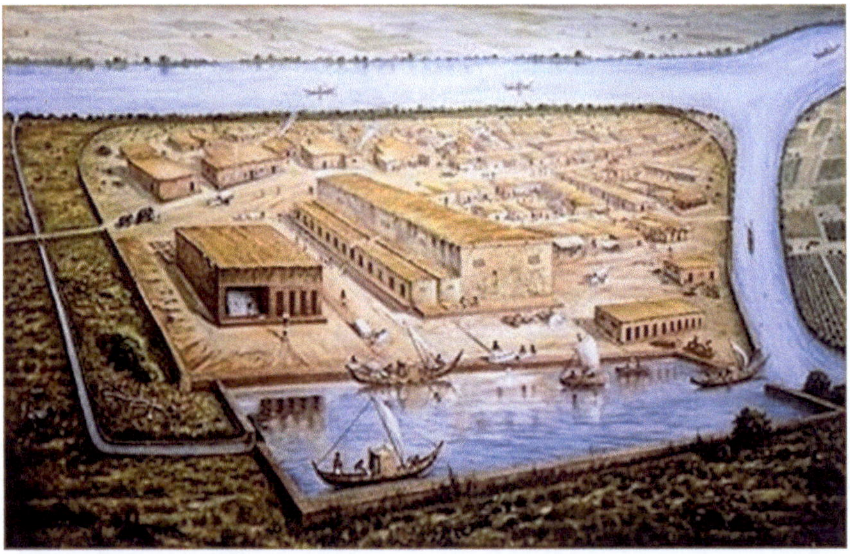

Plate 6.5 Port town Lothal (A conceptual portrayal of S.R. Rao's idea)

Staple Food and Agriculture

The sediments of the Didwānā and Lunkaransar lakes in western Rājasthān contain charcoal, ceramic pieces and burnt stubbles of cereal plants along with pollen grains —the charcoal giving a date between 9400 and 8000 yr B.P. (G. Singh et al., 1974). This means that the people of the Didwānā–Lunkaransar area in the early Holocene time were possibly growing cereal plants.

Remains of two species of wheat and two species of barley were found at Merhgarh of the time span 9000–5500 yr B.P. (Jarrige and Meadow 1998). Rice was a part of food at Harappa 4400 years ago. Earthen pots at Surkotada revealed charred lumps of millet along with the nut *Scirpus* and *Amaranthus*. (Ramarkrishna Rao 2008). At Lothal and Rangpur grains of rice and pearl millet (*bājarā*) were unearthed from the 3000-year old horizon of soil.

As already stated, the presence of the remains of rice have been detected in the Early Harappan Period at Kunāl and Banāwali and of cotton at Mohenjodaro, Harappā and Kunāl. In eastern Kachchh, the people of that time harvested both the winter *rabi*, and summer *kharif* crops, as borne out by findings of remains of barely, wheat, linseed, gram and of rice, *jwār* millet, pearl millet, seasame with cotton (Pokharia et al. 2011).

Occurrence of small-grained millets over a vast swathe of the Harappa land is an eloquent commentary on the pattern of cropping in that time. Sorghum millet and little millet have been recorded from the Early Harappan level at Kunāl, from Mature Harappan level at Banāwali and Lothal, and from Late Harappan sites at Mahorana, Hulās, Sanghol and Pirak, and finger millet and Itlalian millet from the Late Harappa Hulās, and Sanghol (Pokharia et al. 2014). It is significant that there is dominance of finger millets during the Mature Harappan time in the peripheral region of the Harappa land, suggesting shift towards drought resistant crops there (Pokharia et al. 2014).

Obviously there was widespread agriculture in the fertile lands in Gujarāt and in the Sindhu–Saraswatī floodplain. There is a confirmation of this surmise in the finding of criss-cross furrow marks made by a plough (Fig. 6.12a) in the outskirts of Kalibangan (Lal 1971, 1979, 2004). Indeed a terracotta model of a plough (Fig. 6.12b) was also found at Banāwali (Bisht 1984). The existence of granaries, presumably to store grains, in the houses of the Harappan settlement at Mohenjodaro leave little doubt about the Harappan of the rural world being agriculturists. Perhaps there was also an agropastoral society side by side with the urban communities.

Growing food crops was not confined to northwestern–western parts of India alone, the people harvested cereal crops such as rice in the Gangā floodplain as well. Rice grains have been found at Koldihwā, Mahadahā and Lahuradewā dated (8000–2000 yr B.P. and 8570–6530 yr B.P., Fig. 6.13) in Uttar Pradesh (Ramakrishna Rao 2008).

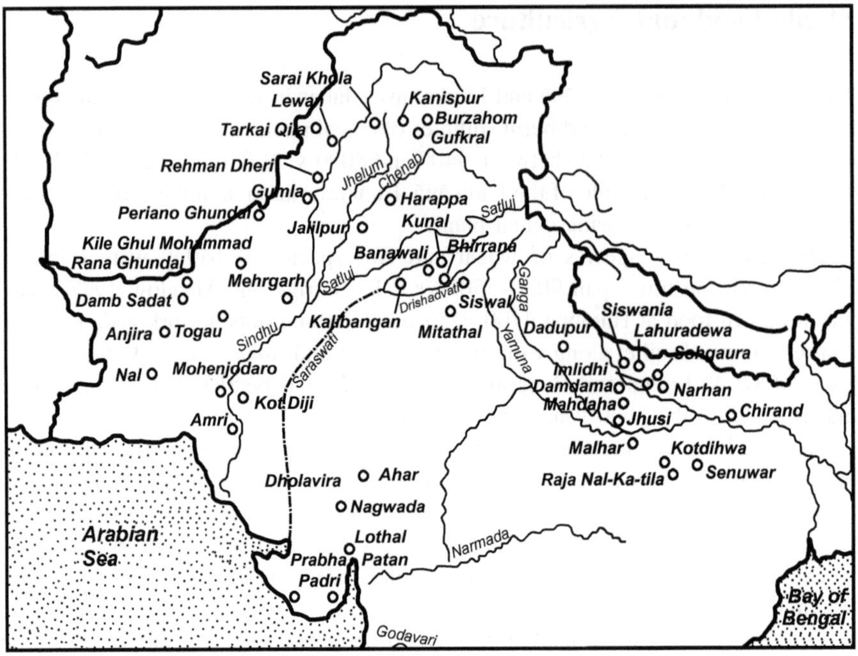

Fig. 6.13 Localities where evidence for cropping cereals have been found in northern India (From Misra 2014)

Characteristic Features of Harappan Culture

There are two features as already stated that stand out as marquee of the Harappan world.

The Harappan cities and towns are devoid of remains of as palaces and memorials Symbols of hero-worshipping are wholly absent. However, there are big structures presumably to house people in power. It seems they had a democratic system of governance and administration which did not distinguish the rulers from the public. There were perhaps city-centric domains bound together in a sort of confederation (Lal 1997). There is no indication of military presence, except that the cities were fortified. There is no evidence of violent conflict and of attack or invasion. It is quite intriguing then why the cities needed fortification.

Vibrant Society

The variety of archaeological material seen in over 2000 settlements across the land of the Sindhu and Saraswatī river eloquently demonstrate that the Harappans lived in a style, had artistic appreciation and elegant tastes. They built brick-made houses having sanitary facilities built in planned cities with well-laid streets and drainage system. The traders used scientific system of weight measurement. And they made ornaments, vessels and tools of copper and bronze. They wore gold ornaments embellished with semi-precious stones. The steatite seals used for business and trade transactions were artistically and meaningfully engraved. A pictorial script (Fig. 6.11d, Plate 6.2) common across the vast expanse of the Harappan world implies that they were intellectually advanced people.

The population of India was then very small and the country was blessed with vast swathes of fertile and well-watered land with rich biodiverse forests and wild life. It is quite natural that the affluent, progressive and vibrant people chose to live for several thousand years since about 5500 years or earlier 9000 years in the domain of the river that was full of flowing water.

The people built their settlements on the limits of the floodways of the river—at some distance from the channel. It is not without significance that the settlements occur along the periphery of the 4–8 km wide alluvial strip of the Ghagghar (Singh et al. 1988). Wise as the Harappans were, they knew the perils of living on the banks of the channel within the floodways, which is prone to recurrent flood. A floodway is that part of the river regime which experiences at least one-foot inundation at least once in a period of 100 years. The floodway conveys the highest flood discharges. This explains why there are no tell-tale evidence of flood-damage in the Harappan settlements.

It may be emphasized once again that the Harappan Civilization progressed in the face of growing aridity in the country—more so in the heartland of the Saraswatī domain. This only shows that the people were capable of coping with hazards of progressively warming climate. And they survived not only the negative effects of climate warming, they remained resilient and enjoyed life and grew in numbers (Valdiya 2013). One would not believe but it is true that the density of human population in this desert at present is highest anywhere in the world deserts. As already stated the human density—57 persons/km^2 in 1961 (Gupta and Parkash 1975). The present (2011) population density is 133 persons/km^2 while the live-stock density is 115/km^2 (Kar 2014b). The reason for high density is obvious: the nutrient-rich alluvial land of Saraswatī continues to provide enough nourishment to natural vegetation and crop plants for the animals and the people to continue to inhabit the land of hostile climatic condition.

Decline of Harappa Culture

Judging from the records of sediments deposited by the Saraswatī in its middle reaches, it emerges that the energetic fluvial sedimentation *prior to* 5000 yr B. P. gradually gave way to a regime of slackened flow. Inherent is the conclusion that the channels were carrying less and less discharge. Simultaneously, the realm of sands of the desert was expanding as testified by the sands of the dunes. However, further downstream active fluvial sedimentation continued until about 2900 yr B.P. when the region was covered by aeolian deposits (Courty 1995).

There is yet another evidence of the decline in the flow of the Saraswatī. The palaeochannels of the river in many areas including Jaisalmer and Pachapadra, degenerated into *ranns*—dry playas made up of finer sediments deposited in standing bodies of water in the fluvial regime (Bakliwal and Grover 1988).

In an environment of growing aridity in the subcontinent, particularly in the realm of the Thār desert, a drastic decline in the discharge of the Saraswatī around 3900 to 3700 yr B.P., brought about radical changes in the lives and the culture of the people of the Saraswatī floodplain. The Late Harappa period (3900–3300 yr B. P.) is indeed the time of the decline of the once glorious civilization. This is discernible in the fall of civic senses of the people in even such progressive cities as Dholāvīrā. They started encroaching on the public space by extending private platforms and pottery kilns onto the sheets. There were no efforts made to clean up reservoirs which had silted up heavily. Material removed or pilfered from older building were used to make new houses; and the bricks made were of poor-quality (Bisht 1991, 1998). New towns that grew up lacked planning and public facilities like the sanitary system.

Ornaments became rare or even absent, presumably due to loss of affluence or absence of raw material. Although seals were used, the overseas trade seems to have declined; and commerce was largely inland. Only the script was used on the seals while the motifs of animals, composite animals or sacred symbols were completely discarded. Moreover, the square shape of seals became rectangular, rather plano-convex. One could see regressive signs in such places as Dholāvīrā, Rangpur and Lothal. However, cultural elements such as ceramics tradition remained alive.

Most important development was the urban culture giving way to predominant agricultural occupation, particularly in the frontier areas of resettled people who were uprooted from the middle Saraswatī floodplains. The new settlements were smaller in size but larger in number. Recent findings at Chandayan in Baraut tahsil, district Bāghpat between the Yamunā and Gangā rivers, reveals houses of mud walls, the houses containing bowls, dishes, flasks and lids with knob and cylindrical agate beads. In the proximity were remains of a human skeleton (femur and pelvis bones), a broken upper crown placed on the skull, and animal bones (A.K. Pande, *The Hindu*, January 3, 2015). The remains suggest funeral ceremony of sorts. Even at Banāwali, Sanghol and Hulās, the houses of this late phase were made of pressed earths, not mud bricks.

With the abandonment of towns and cities around 3750 yr B.P. (Lal and Thapar 1967), the fabric of the civic system degenerated considerably in the few settlements that were still inhabited. The decay of the culture seems to be related to environmental stresses that the death of the once benevolent river brought about. This happened in the face of aridity that prevailed over the entire Sindhu–Saraswatī land. A resilient culture that not only withstood the stresses of aridity but flourished despite dryness over a thousand years (of the Mature Harappan Period), finally crumbled when the life-giving Saraswatī itself became lifeless.

Availability of iron and its use for making tools 3200–3100 years ago facilitated clearing of forests and reclaiming land for agriculture. Wood being aplenty in the land where eco-refugees had resettled, the pottery making and brick-manufacturing industries saw considerable advances. Before that happened, the late phase of the Indus tradition is variously known as Cemetery-H, Jhukar and Bara. This was followed by the Early Iron Age cultures represented by the Painted Grey Ware, and the Black and Red Ware cultures in different parts of the country (Fig. 6.14). The Ochre-Coloured Pottery Culture (3700–3300 yr B.P.) was coeval with the Late Harappan Culture of the Saraswatī–Sindhu domain.

During the Painted Grey Ware Culture (3200–2500 yr B.P.) agricultural economy progressed, the people learnt iron technology and craft and the industry saw specialization (Bhan 2000). It is this culture which is believed to have figured in the narratives of the epic *Mahābhārat*. Archaeological excavation at Ālamgīrpur on the bank of Hindan river and at Hastinapur (District Meerut, U.P), Ropar (Panjāb), Sanghol (District Fatehgarh Sahib, Panjab), revealed evidence of the Painted Grey Ware and Northern Black Polished Ware cultures (and of the Sung-Kushan and Early Mughal cultures) (Lal 2002). Among the artefacts found were cubicle dices marked with 1–6 circles on the six facets of the cube.

New Developments in Gujarāt Coastal Belt

The influx of people uprooted from the lower reaches of the Saraswatī. spurred the growth and development of cities in the coastal belts of Kachchh and Saurāshtra—particularly where there were seaports. Among others, Dwārkā was rebuilt and fortified with circular and semi-circular bastions.

Prabhāspattan and Lothal continued to have overseas trades. An incredibly large dockyard (Plates 6.3 and 6.5), larger than that of Visākhapatnam (Mathur 2002)—along with a huge warehouse to stock export-import goods at Lothal (Rao 1991) speak volume of the vigorous overseas trade. In the port city Khirsara on the Khari river in Kachchh (the Rann then was under sea water) there was a multipurpose warehouse 28 m long and 12 m wide with 14 parallel walls, each 11 m in height, for storing goods for export or import (R.N. Kumaran, *The Hindu*, August 6, 2013).

The coastal India was thus a land of progress and industry.

The cultural deterioration is writ large in every aspect of life of the people that inhabited the land. It was the twilight of the glorious Harappa Civilization. The culturally sophisticated and socially upward-mobile civilization was eventually replaced by a multiplicity of regional culture, predominantly rural in economy.

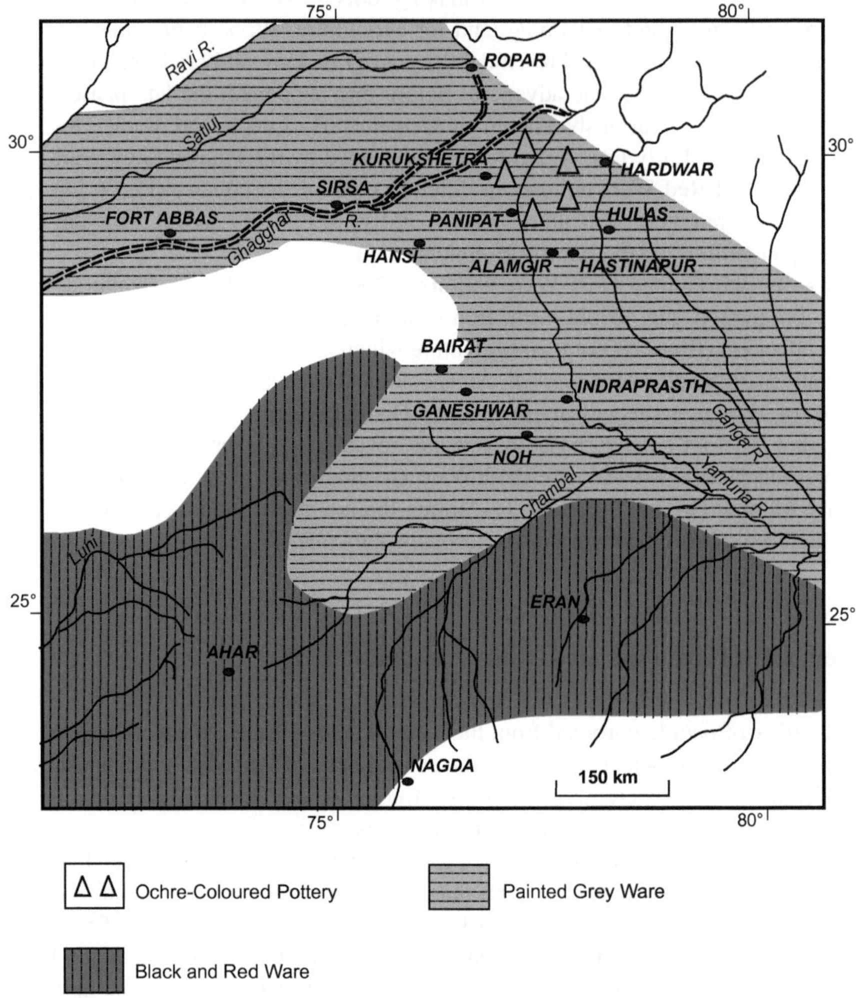

Fig. 6.14 Three of the five post-Harappan cultures in northern India (Based on Thapar 1979; Alchin and Alchin 1993). These cultures are based on development of pottery craft

Chapter 7
Decline and Demise of Saraswatī River

Decline of Discharge

A comprehensive sedimentological study carried out in the middle reaches of the Ghagghar between Tohana and Sirsā by Courty (1995) unfolded a history of progressive decline in discharge and reduction of sediment deposition in the period 7000–5000 yr B.P. Concurrently there was increase in wind activity. This period was followed by silting up of river channel and formation of localized dunes 5000–2500 years ago. Occurrence of black clay horizons within the column of sand and silts (Fig. 5.6) imply development of marshes and lakes in the riverine regimes resulting from stagnation of water or ponding of the river. The clays and muds are characterized by carbonate concretions, the hallmark of aridity that must have then become prevalent with occasional spells of rainfall. Even as the Thār desert expanded, the aeolian sands covered the floodplains of the Saraswatī.

The OSL dating of sediment demonstrates that the Saraswatī that was very active in the period 5900–4300 yr B.P. became very sluggish in the period 4300–2900 years ago (Saini et al. 2009). Lack of incision in the wide valley within the Ghagghar–Hākrā interfluves, according to Giosan et al. (2012) implies that in the period 4300–2900 yr B.P. the Saraswati was no longer a major river having substantial discharge. However this is partly true. For, in the Tohanā–Sirsā reach, the non-meandering Ghagghar is entrenched—the 30–100 m wide valley is incised 3–6 m deep (Saini et al. 2009). There must have been more prominent incisions before the floodplain was covered by the fine sediments brought by dust storms blowing for over 3000 years. Neotectonic activities including earthquakes also obliterated the earlier formed landscape in the upper reaches of the river (Valdiya 2013).

If one were to believe in the credibility of the narratives in the epic *Mahābhārat* authored sometimes after 3500 yr B.P. by Krishna Dwaipāyan 'Vyās' (Valdiya 2012), the Saraswatī had lost much of its water during the *Mahābhārat* time (before 3500 yr B.P.). The Pāndavs during their exile saw the river disappearing under the sands (of the desert) at Vinashan (*Mahābhārat,* Van Parv, 25)—*Vinashanam rājan*

K.S. Valdiya, *Prehistoric River Saraswati, Western India*,
Society of Earth Scientists Series, DOI 10.1007/978-3-319-44224-2_7

jagāmāth halāyuddhah. Yatra nashtā Saraswati. Krishna's elder brother Balarām in his pilgrimage from its mouth at the sea to its headwaters in the foothills along the course of the Saraswatī had also seen the river drying at Vinashan. Downstream of Vinashan, Balarām had noticed the *antahsalilā* (underground river) river Saraswatī reappearing at a number of spots such as Chamsodbhed, Nagodbhed, Shivodbhed and Udapan where centres of pilgrimage had sprung up. Balarām was able to recognize the underground Saraswatī by detecting moisture in the soil bearing a carpet of vegetal greenery (*Mahābhārat*, Shalya Parv 35, verse 90)—*snighdhatvād adoshdheenām cha bhumeshcha Janamejaya jānanti siddhā rājendrā rahtāmapi Sarasvateem.*

On the basis of the astronomical data such as constellation of stars and solar-lunar eclipses, 64 astronomers dated the Mahābhārat War at about 5000 yr B. P. and 42 astronomers, including Ashok Bhatnagar (2014) fixed its time around 3500 yr B.P. Iyengar (2003) places the date around 3478 yr B.P. On the basis of genealogy of the descendants of king Parikshit (grandson of the Pāndav warrior Arjun) the time of the Mahābhārat War is estimated at about 3400 yr B.P., (Mallya 2015). There is yet another indicator, albeit indirect, of the time of the War. Bet Dwārka, the satellite town of Dwārka, is protected against the ocean by a stonewall. Sunken into the sea, this wall contains pottery pieces embedded with rocks. S.R. Rao had recovered, besides the retaining wall, the Harappan seals along with a house wall and profusion of Harappan pottery. Thermoluminiscence date of the pottery pieces is 3528 yr B.P. (Rao 1999). In the Mahābhārat (Mausal Parv V, Chap. 7), it is stated that 36th year after the Mahābhārat War (Ganguli 1998) the city of Dwārka was swallowed by the sea—that is sunk into the sea. That places the event of War around 3500 yr B.P.

The above narrative finds strong support from the deduction arrived at by archaeologists. The archaeological evidence points to large-scale exodus of the people at about 3750 yr B.P. (Fig. 7.1) from the middle reaches of the Saraswatī (Thapar 1975), presumably triggered by drastic reduction in the volume of water flowing through the channel of the Saraswatī. The exodus happened probably much earlier around 4000 yr B.P. (R.S. Bisht, personal communication, 2015). The sudden and substantial decrease in the river discharge forced the people to leave their hearths and homes and resettle in greener pastures elsewhere—in the upper reaches of the Saraswatī basin such as Mandā, Ropar and Chandīgarh and in the adjoining parts in northwestern Uttar Pradesh—in the Yamunā–Gangā interfluves (Fig. 7.1). In addition to Ālamgīrpur on the banks of Hindan River (Meerut district), the other notable places are Hulās, Ambikheri, Krishni, Chilkhera and Nayābans—the concentration of settlements being in the Sahāranpur district (Dixit 1981, 1993). The development implies a geological event that happened swiftly and severely, bringing dramatic changes in the drainage. It must have been the tectonic upheaval, entailing fissuring and attendant sinking and uplift of the land through which the eastern branch of the Saraswatī (the Tamasā River) flowed (Valdiya 1998, 2002). As already detailed (Chap. 3), the Siwālik subprovince and the foothills belt have been repeatedly affected by tectonic turmoil in the Later Pleistocene to Holocene time. Quite a few faults and thrusts had become active,

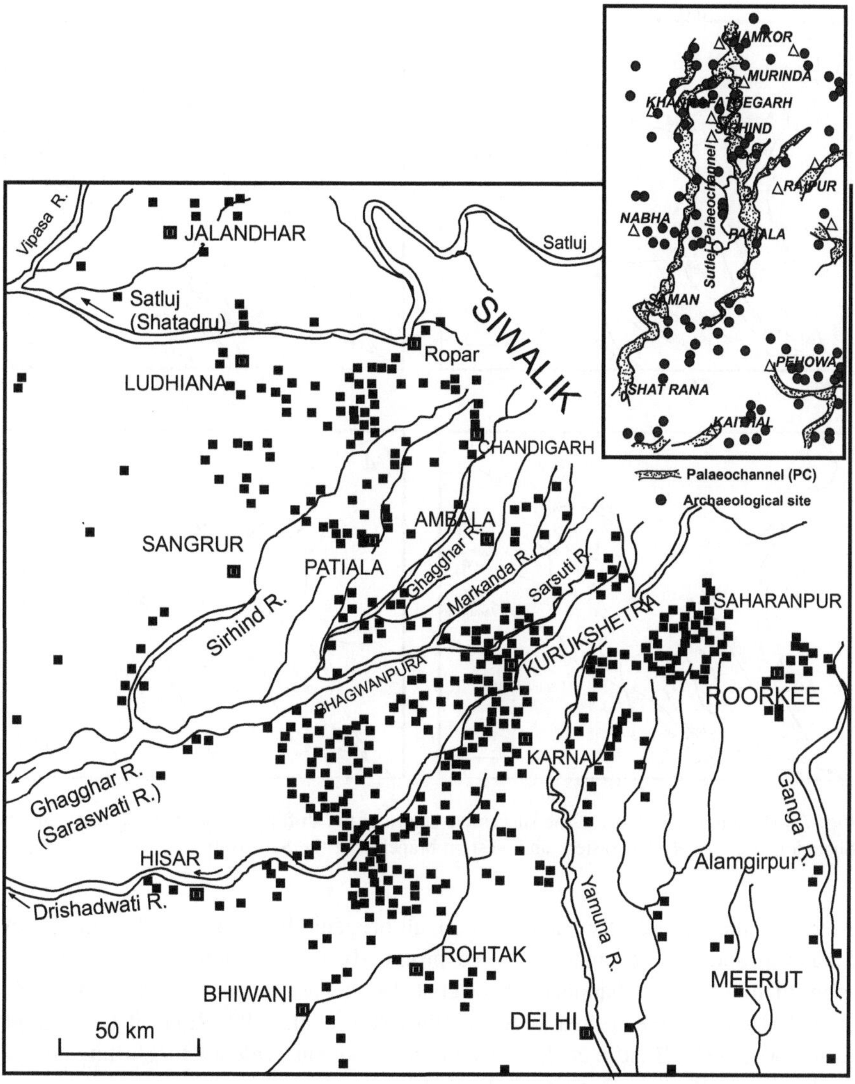

Fig. 7.1 Suddenly a large number of Later Harappan settlements appeared in the foothills of the Siwālik—in the upper reaches of the Saraswatī basin and in the adjoining northwestern Uttar Pradesh. This happened following large-scale exodus of the people from the middle reaches of the Saraswatī (Based on Joshi and Bisht 1994). Inset Late Harappan settlements on the banks of the channels through which once flowed the river Shatadru (Satluj) before it swerved west (After Sharma and Bhadra 2012)

registering sideways (strike-slip) and up-and-down (dip-slip) movements again and again. As a consequence, some parts of land sank, others rose up, and some other parts were dislocated. Under this condition the streams and rivers were forced to change their courses; some were deflected and some others were robbed of their

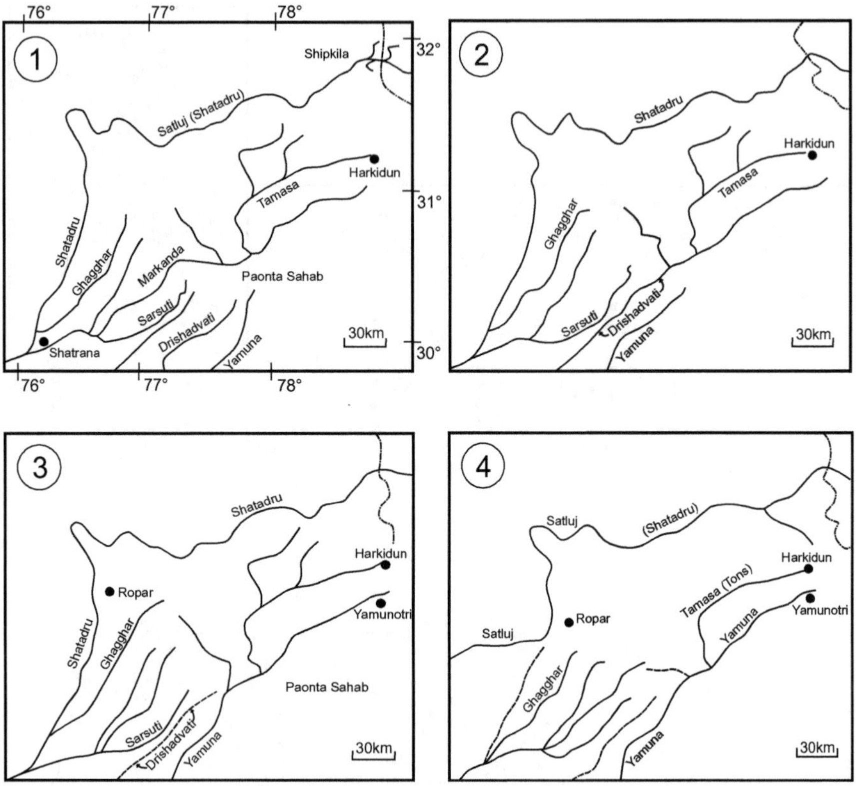

Fig. 7.2 Digrammatic maps of the successive stages of the shifting of the rivers Tamasā and the Shatadru, respectively the eastern and western branches of the Saraswatī

water. As the tectonically resurgent Arāvali orogenic belt very slowly rose up, the Saraswatī and their tributaries shifted progressively westwards. The existence of multiplicity of palaeochannels (Ghose et al. 1979; Yashpal et al. 1980; Ramaswamy et al. 1991; Kar 1988, 1999a, b; Sahai 1999; Nair et al. 1999; Rajawat 1999; Gupta et al. 2004, 2008; Bhadra et al. 2009) provide eloquent evidence for shifting courses of the Saraswatī and other rivers (Plate 4.1).

Deflection of Eastern Branch of Saraswatī

The abrupt exodus of the people of the middle reaches of the Saraswatī around 3750 year ago (or earlier around 4000 years B.P.) implies, as already stated, that there was an event that happened suddenly and swiftly. This is further borne out by sudden appearance of a large number of settlements in the upstream foothills belt. This event is attributed to southwestward deflection of the eastern Tamasā branch of

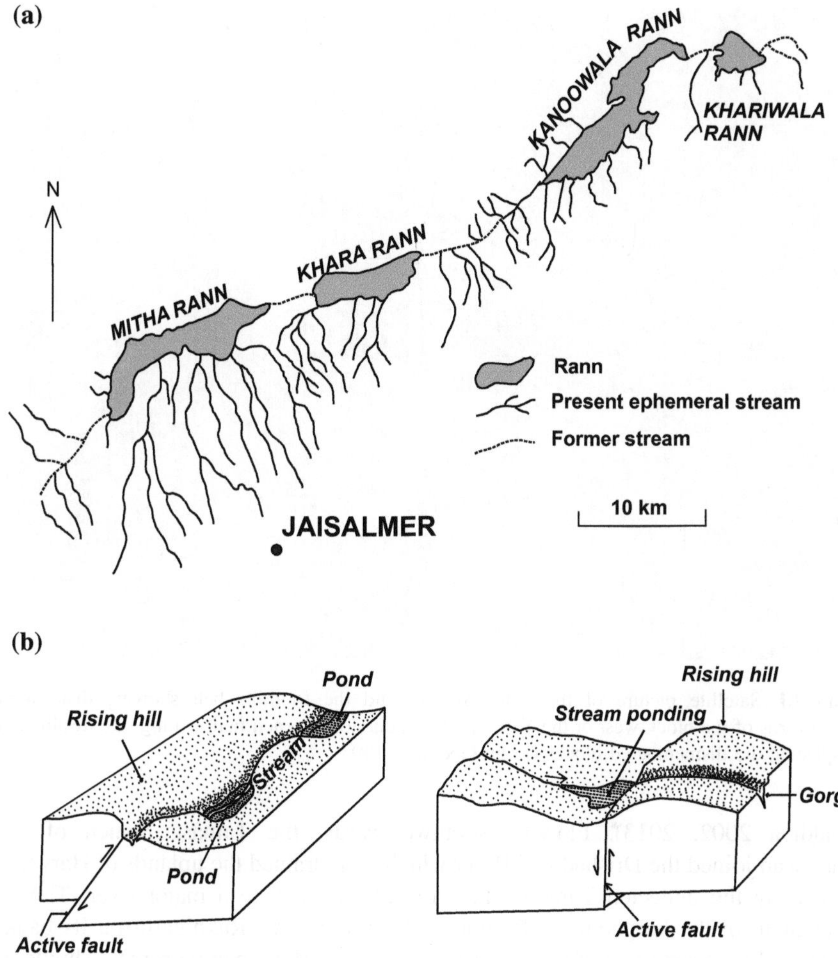

Fig. 7.3 **a** Streams are segmented resulting in the formation of a chain of lakes (Roy 1999). **b** Block diagrams explain how movements on active faults that cross the streams or follow their courses bring about their ponding and resultant formation of lakes (From Valdiya 2002)

the Saraswatī (Oldham 1893; Valdiya 1968, 1998, 2002, 2013). This happened as a consequence of a NNW–SSE trending wrench fault, the Pāontā Sāhab Fault, or the Yamunā Tear fault, wrenching apart the outer range of the Siwālik, displacing the downstream western block laterally southwards, and simultaneously uplifting it by nearly 20 m (Srivastava et al., 2013; Plate 7.1; Figs. 7.2 and 3.3).

The uplifted and laterally displaced block thus impeding its flow, the Tamasā found the highly crushed and weakened rocks of the fault zone quite easy to cut and make a new channel for itself. It flowed through this newly formed channel

Plate 7.1 Satellite picture of the outer Siwālik and the foothills belt showing dislocation southwards of the block west of the Yamunā River, due to a wrench fault tearing the Siwālik and displacing the dismembered blocks (From Google Earth)

(Valdiya 2002, 2013). Flowing southwestwards, the Tamasā branch of the Saraswatī joined the Drishadwati River which then drained the uplands of Haryānā. Joined by the deflected Tamasā, the Drishadwati became a major river. This is evident from the burgeoning of among others, the town Rākhigarhi on its bank. Rākhigarhi, became one of the biggest towns of the Harappan world (V. Shinde in Subramanian and Krishnan 2014). The fluvial deposits of this river bear testimony to its being a great river.

This must have happened sometime after the temporal interval 3878 yr B. P. ($\leq$ 3663 $\pm$ 215 yr B.P.) when not far from the Pāontā Sāhab Fault (also called Yamunā Tear), a stream-gravel deposit (Shāhjahānpur) containing 3663 $\pm$ 215 yr old carbonaceous matter in the foothills was lifted up 20–30 m as a result of movement on the active Himālayan Frontal Thrust (Wesnousky et al. 1999). At Trybron village close to the HFT and in the proximity of the Pāontā Fault, a 9-m high scarp on the gravel deposit was formed, which forced the capture and deflection of Nagalkhol stream (Wesnousky et al. 1999). The HFT must have been reactivated due to faulting along the transverse wrench fault—the Pāontā Sāhab Fault. Admittedly one single data cannot be taken as a very creditable datum to build a big picture. Doubtless, more dates are needed to establish the time of the river swinging southward. Sometimes later the land in the *foothill* to the east of the

NNW–SSE Pāontā Sāhab Fault (Fig. 3.3; Plate 7.1) sank 14–30 m (Thussu 1999). The sinking of the ground forced the then southwestward flowing river to deflect southwards towards the then Yamunā, (Fig. 7.2(3)) which was a tributary of the Chambal River—itself a tributary of the Gangā. The Tamasā branch of the Saraswatī now joined the Yamunā, which thus became enriched considerably. *The Yamunā became the carrier of the eastern branch of the Saraswatī water and delivered it to the Gangā at Prayāg (Allāhābād). Earlier, Prayāg was the confluence of the Gangā and the Yamunā. Now the Saraswatī also joined it through the channel of the Yamunā. The Prayāg became the Trivenī.*

Subscribing to the commonly held view that the Saraswati originated in the Himālaya, Danino (2010) suggests that only a portion of the Yamunā–Tons flowed westwards into the Mārkandā valley, the rest flowing southwards through the opening formed along the Yamunā Tear Fault. The westward branch was the Saraswatī and the southward branch the Yamunā. And when it debouched into the plains, the Yamunā divided once again—a part flowing southwestwards and joining minor streams of the Drishadwati, the other flowing south as the Yamunā Later, the Eastern Yamunā captured most of the water, thus completing depletion of both the Drishadwati and the Mārkandā.

The main Saraswatī in its heartland was now left with only the waters of the Shatadru, its western branch.

Demise of the Saraswatī

To recapitulate, the foothill belt of the Saraswatī basin was in tectonic turmoil all through the Later Quaternary times. In the Later Holocene the WNW–ESE oriented Lahore–Sargodha subsurface ridge under the Panjāb alluvial plain was affected by tectonic movements. Possibly a spasmodic movement like uplift of this under-surface ridge caused the subsidence of the ground to its west, forcing the Shatadru to swerve abruptly westwards from its earlier southerly course (Valdiya 1998). The suddenness of this development is manifest in the sharp U-turn the river makes at Ropar (Fig. 7.2(4)). Thus the Shatadru (Satluj) abandoned the Saraswatī and joined the Vipāsh (Beās) river, a tributary of the Sindhu. This happened around 2600 yr B.P., for there was another exodus of the people from the Cholistān area (Mughal 1995).

The Saraswatī fell victim to river piracy for the second time. The later phenomenon deprived the river of its entire discharge. This development as already stated, triggered another wave of exodus from the Hākrā reach in Cholistān (Mughal 1995). This time the people migrated southwards to Gujarāt and beyond. Some went far south to settle down along the West Coast where they are now known as *Sāraswat*. Twice robbed, the Saraswatī was reduced to a petty rivulet - a misfit river - moving in an uncommonly wide and sand-filled dry channel. The land once watered by a many splendored mighty river became a parched dry land, the western part of which was encroached on by the moving sands of expanding Thār

Desert. The arid tract is now characterized by a number of saline lakes and flat-bottomed basins in the desert occupied by ephemeral lakes called playas forming a curved chain (Fig. 7.3a). The lakes were formed when during intense aridity the channel of drying river was divided into segments by movements on active faults that crossed the river. The segmentation of the channel resulted from lateral displacement, uplift and sinking of the fault blocks (Roy 1999; Roy and Jakhar 2001). The movements on faults caused blockage, resulting in the formation of lakes within the riverine regime (Fig. 7.3b). Gradually drying up, these lakes turned saline, and eventually became playas.

The life of the once 'finest of all rivers' 'surpassing all rivers in splendour and magnificence (*Rigved*, 2:41:16; 6-61-2; 6-61-81) thus ended in a tragedy of complete loss of water.

But then the Saraswatī could not have been wholly waterless—at least in its middle reaches. For, historical records show that the Satluj, a Himālayan river all through its long life—was flowing into the Ghagghar until about 1796 CE (*Imperial Gazetteer of India*—1908).

The foregoing narrative presents only one viewpoint, namely tectonically forced river-capture in a land riddled with active faults and frequently affected by neotectonic movements. As rightly pointed by Giosan et al. (2012), this theory lacks very precise chronologies reconstructed from well-documented outcrops and/or cores. The alternative theory to explain the decline and demise of the Saraswatī needs to be examined.

Weakening of Monsoon in the Saraswatī–Sabarmatī Land

A number of researches in recent years subscribe to the notion that reduction in monsoon rain is responsible for the death of the River Saraswatī.

On the basis of comprehensive study entailing use of Shuttle Radar Topography Mission data combined with fieldwork and radiometric dating, Giosan et al. (2012) concluded that climate change provides a full and relatively well-supported explanation. In the region where aridification intensified during the Late Holocene, the weakening of monsoon caused the gradual drying up of the monsoonal-fed river. The main plank of their argument is that the Saraswatī was not a glacial-fed Himālayan river, but a monsoonal rain-fed foothills river; that upstream of its alluvial plains there is a lack of large-scale incisions, and that downstream sedimentation slowed down on the distinctive megafluve ridge simply because of 'fluvial quiescence' resulting from gradual decrease in flood intensity.

A variety of geological evidence shows without ambiguity that the period of high rainfall in the temporal span 8500–3500 yr B.P (Singh et al. 1974; Bryson and Swain 1981; Wasson 1996) was followed by a time of aridity at 3500 yr B. P. according to Naidu (1995). In the region encompassing the basins of the Saraswati and the Sindhu there was intense aridity around 4200 yr B. P. (Staubwasser et al. 2003).

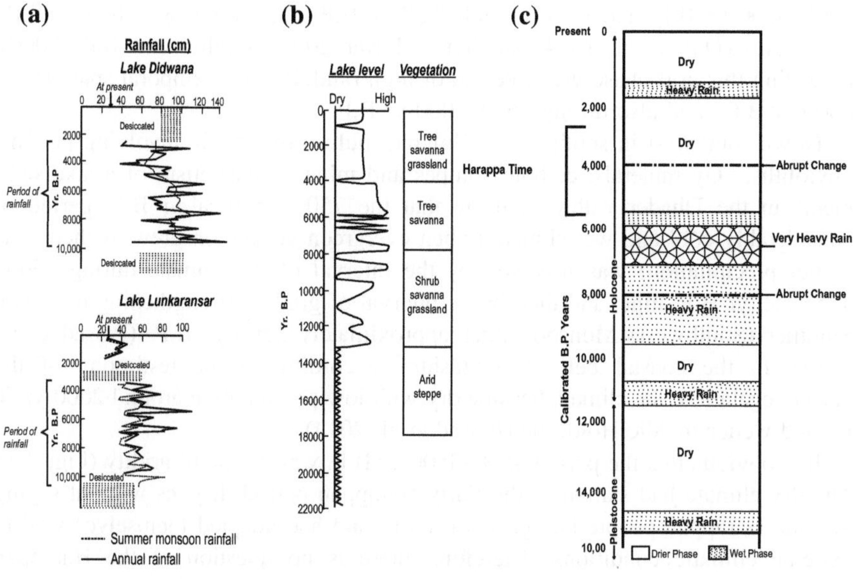

Fig. 7.4 **a** Pollens recovered from the sediments of the Dīdwānā and Lunkaransar lakes testify to spells of rainfall in an otherwise dry climate (After Wasson 1995). **b** There was prolific growth of plants whose pollens lie buried in the sediments of these lakes (After Bryson and Swain 1984). **c** Pattern of rainfall over the Indian subcontinent in the last 15,000 years. Notice the alternating dry and wet climate (After Singhvi and Kale 2009)

A large body of proxy evidence clearly indicates that in the time of the Mature Harappan Period (4600–3900 yr B.P.), the climate was dry with reduced rainfall from Indian Summer (Southwest) Monsoon all over the Indian subcontinent, including the basins of the Saraswatī, the Sindhu, and the Sabarmatī systems.

Not far away from the upper-middle reaches of the Saraswatī the sediments of the Kotla Dahar palaeolake in Mewāt district, southern Haryānā, provide evidence of the weakening of Indian Summer Monsoon at ~4100 yr B.P. as manifest in peaking up of evaporation–precipitation ratios (Dixit et al. 2014). High resolution oxygen isotope study of sediments of the Lunkaransar Lake within the parched domain of the middle reaches of the Saraswatī, shows that the water level rose to the maximum around 6300 yr B.P. Thereafter, there was abrupt decline and the lake dried up by 4800 yr B.P. (Enzel et al. 1999). This implies that the people of the Mature Harappan Period lived and flourished several hundred years after the desiccation of the lake due to onset of aridity.

Study of pollens from the playa sediments in the Bāp, Maler and Kanod areas in Western Rājasthān shows that before the Mature Harappan Period, the 8500–5500 yr B.P. temporal span was the time of appreciable rainfall when vegetation grew prolifically (Deotare et al. 2004). Things were different afterwards. The calcareous concretions in the sands and soils of the Thār Desert provide credible testimony of intense aridity (when dry winds blew in cyclic fashion) during the time

5000–3500 yr B.P. and then around 2000 yr B.P. and during 800–600 yr B.P., (Fig. 7.4c) (Dhir et al. 1994; Singhvi and Kar 2004; Singhvi and Kale 2009). Preceding this arid phase was a period of high rainfall in the temporal span 9500–6500 yr B.P., and also during 2600–2000 yr B.P.

Down southwest in southeastern Gujarāt, multi-proxy study involving pollens, phytoliths, clay minerals, carbon isotopes and mineral magnetism of a lake sediments in the Dhader valley, indicate that the 7500–5560 cal yr B.P. period of considerably cool and wet climate (when evergreen vegetation grew as there was winter precipitation) was followed by the interval of dry climate during 5560–4255 cal. yr B.P. time, and then by an interval of gradual strengthening of Indian Summer (Southwest) Monsoon after approximately 3500 yr B.P. (Prasad et al. 2014). In the coastal belt of Saurāshtra, according to the testimony of the palaeovegetation, the climate became dry with low precipitation around 2000 yr B. P., and wetter middle Holocene (Prasad et al. 2014).

It is obvious that the period 4800–3500 yr B.P. was a time of aridity (Fig. 7.4). The dry climate had started in the Early Harappan period. It goes without saying that the Harappan people had grown used to and had adapted themselves well to extreme climatic conditions. Therefore, there is no question of the Harappan Civilization collapsing or meeting its demise (Valdiya 2013).

Not only the Sindhu–Saraswatī land but also the whole of India (Table 7.1) experienced decline in rainfall around 3500 yr B.P. (see Valdiya 1999a for references). As a matter of fact there was continental-scale decline in the precipitation from the Summer Southwest Monsoon (and possibly the Westerly rain) all over Central and South Asia (Wright et al. 2008; Giosan et al. 2012). It seems logical to surmise that this continental-scale aridity—which must have been very severe in the

Table 7.1 Climate change during the Holocene in the Indian subcontinent (From Valdiya 2002)

Area (Testimony of)	Warm–wet (yr BP)*	Hot–dry (yr BP)*
Arabian Sea (Planktonic foraminifers)	10,500–5000	3500
Western Coastal Belt (Terrestrial detritus)	10,500–10,000	
Kārwar Coast, Karnātaka (Land-derived pollens)	10,500–5000	3500
Nīlgiri Hills, Tamil Nādu (Carbon isotope values of peats)	9000–8000	5000–2000
Dīdwānā–Lunkaransar Lakes, Rājasthān (Pollen profiles)	10,800–4500 (Peak 6000)	4000–Present
Gangotri Glacier, Garhwāl (Pollens in sediments)	6500–4000	3500–3000
Tsokar Lake, Ladākh (*Juniperus* flora)	10,000	
BangongCo Lake, Ladākh (Pollens)	6000 (maximum)	4000–3000
South-eastern Tibet (Pollens in lakes)	7500–3000	3000–1500
Bhimtāl Basin, Kumāun (Pollens)		3550 ± 120
Rārā Lake, West Nepāl (Pollens)		4500
Karewā Lake, Kashmīr (Pollens)		5000

*year BP—years before present

desert land—brought considerable reduction in the discharge of the River Saraswatī, particularly in the period 4000–3500 years B.P.

The climate being so dry, so rainless, it is difficult to explain the presence of bones of such water-loving animals as elephants, one-horned rhinos and water buffaloes at Kālibangan as well as in some places in Gujarāt (Nath 1969; Banerjee and Chakraborty 1973). Either the bones were exotic — brought there from rain forests which were far-far away, or that water continued to flow down the channels of the rivers of the land. There is another indicator, albeit indirect, of the presence of water in the channels. The Harappan settlements in the Saraswatī domain are restricted to rather narrow tract on the two sides of the major water courses, implying that the people depended heavily on the river water in the atmosphere of prevalent aridity.

There is one more point that stands out against the postulation that the reduction of monsoon rains caused wholesale drying up of the Saraswatī. Since the aridity prevailed all over the Indian subcontinent, including the Himālayan heights, rivers other than the Saraswatī that rise from the Siwālik hills and their foothills (such as the Hindan, immediately east of the Yamunā; and the Gomati, east of the Gangā among others) should have also dried up or become monsoonal-fed seasonal when the rainfalls dwindled. However, these rivers and scores of other rivers of similar origin continued to flow despite the prevalence of dry climate. The Gomati River originates from the lake Fulhaar, 3 km east of Pilibhit about 50 km away from the Siwālik Hills. The river supports life and agriculture throughout the year all along its reaches up to the point of its meeting the Gangā in the Varanasi district.

It is therefore doubtful that rainfall decline alone is the cause of the demise of the Saraswatī. There are, and have been, other factors, some more powerful, which brought about the dismal development that overtook the Saraswatī River as already explained. Explanations are needed not just for the drying up of the river but also for shifting of river courses, river captures, abrupt stream behaviours and sudden exodus *en masse* of the people from their centuries-old preferred habitations.

Precipitation in Himalayan Catchment of the Saraswati River

More important than the condition of climate and rainfall in the populated Saraswatī floodplain, are the behaviour and the trend of precipitation in the catchments of the two major feeders of the Saraswatī in the Himālayan province from where they emerged. It is the intensity and duration of rainfall *and the amount of snowfall in the catchments* of the Shatadru (Satluj) and the Tamasā (Tons) that matter and are of supreme, rather critical importance.

As of the present, the Satluj–Tons sector of the Himālaya happens to be in the zone which is affected by both the *Indian Summer Monsoon* (also called *Southwest Monsoon*) rising in the Indian Ocean and the *Winter Westerlies* (Plate 7.2) which

Plate 7.2 Satellite picture shows the trend and the range of the Winter Westerlies (From Google Earth)

come from the Mediterranean Sea. *This transition zone—in which the Satluj and Tons emerge and flow through—is thus the beneficiary of the two weather systems.*

There is no reason to doubt the surmise that this climatically sensitive zone is for millennia receiving precipitation round the year in the form of rainfalls and snowfalls. The more intense and prolonged the Westerlies, the greater would be the volume of snow and ice in the Himālayan catchments. Indeed, the "Western disturbances (Plate 7.2) are extra-tropical synoplic-scale weather systems which cause significant precipitation in the Himālaya and surrounding areas during winter" (Dimri and Chevuturi 2014).

The snow and ice melt in summer. The warmer the summer, the larger the quantity of melt water and the greater the volume of discharge in the rivers. It is therefore logical to conclude that despite reduced rainfall from the Indian Summer Monsoon, the Himālayan-born Saraswatī must have been receiving melt-water when the climate turned arid in the floodplain. The melt water that flowed down the multiple channels of the Saraswatī was sufficient to support the life in this floodplain.

A study showed that water in the neighbouring Beās River did increase around 5500 yr B.P. until about 4100 yr B.P. (Wright et al. 2008). According to Jaishri Sanwal (Sanwal et al. 2013) "*the Himālayan region received greater rainfall than the other parts of India during the Late Holocene*" *due to stronger Westerlies*, "and not entirely because of the decrease in the Indian Summer Monsoon in the Peninsular India". Investigation involving combined oxygen and carbon isotope analysis of sediments of palaeolakes and existing lakes *in the Central Himālaya (embracing eastern Himāchal Pradesh and Uttarākhand), unambiguously demonstrate that in the last 4000 years (Late Holocene) the precipitation was from two different systems* (Kotlia et al. 2014). "*The* $\delta^{18}O$ *values show strong variability from* −2.1 *to* −8.9 ‰, *pointing to precipitation from two different sources—the Westerlies playing an important role in the Late Holocene climate of the Indian Himālaya*" (Kotlia et al., 2014).

Not only the Himālaya but even the vast plain of the Saraswatī in front of the central Himālaya, must have received rain from the Winter Westerlies, as is happening in the Panjāb and Haryāna in the present. Specific examples of Late Holocene rain and snow are given below.

Variation in Precipitation Trend in Central Himālaya

Multi-proxy studies using pollens, clay minerals, mineral magnetism, carbon isotopes, etc., from palaeolakes and existing speleothems from caves in eastern Himāchal Pradesh and western Uttarākhand and Western Tibet that encompass the catchments of the Satluj (Shatadru) and the Tons (Tamasā) throw revealing light on the trend of precipitation in the last 25,000 years or more. It emerges that the periods of dry and cold conditions have alternated time and again with the period of wetness, warmth and humidity (in terms of Indian Summer Monsoon). The duration of the periods have not been the same and also not synchronous—everywhere. They varied from basin to basin. One thing is, however, common. During the periods of dry and cold condition evergreen forests (such as of *Quercus* oaks, aldar) and grasses dominated the vegetation while the wet, humid and warm phases witnessed growth of deciduous forests, including conifers. The presence of evergreen vegetation in cold–dry times implies availability of adequate moisture in the soil, which only precipitation either in the form of rain or snowfall would have ensured and sustained. *Despite the reduction of rainfall from the Indian Summer Monsoon there was significant amount of precipitation from Winter Westerlies, enough to support vegetation and enough to replenish river discharges.*

To build a case for the Saraswatī River having adequate water discharge during the Mature Harappan Period (4600–3900 yr B.P.) in the land afflicted with aridity, it will be worthwhile to dwell on precipitation record of only the limited period from about 5000 yr B.P. until about 3000 yr B.P.

The Spītī River in Kinnaur in northeastern Himāchal Pradesh is the most important tributary of the Satluj (Shatadru) in its upper reaches. Palaeolakes in the river regime provide a record of climate and rainfall. Despite reduced monsoonal precipitation from Indian Summer Monsoon, cold dry climate prevailed in the period 4800–2000 yr B.P. in this region and glaciers advanced in the interval 3000–1500 yr B.P. (Owen et al. 2005). The implication is obvious: the Spītī valley was under the influence of Winter Westerlies. In the upper reaches of the Lāhaul valley (to the northwest of the Spītī and beyond the Baralacha Pass) the Yunom-river palaeolake testifies to prevalence of cold dry condition in the time span 4500–2000 yr B.P. (Bohra and Kotlia 2014). Analysis of carbon isotopes, total organic carbon and pollens in the peats and sediments of a lake in the Chandrā River in Lāhaul demonstrates that warm and wet climate promoting prolific growth of broad-leaf trees as well as alpine-meadow-vegetation in the period 6732–3337 cal yr B.P. was followed by a spell of cold dry climate from 4808 cal yr B. P. to 4327 yr cal yr B.P. (Rawat et al. 2015). The period 3337–2032 cal yr B. P. was the time of warm and wet climate.

Coming to the Tamasā (Tons) catchment encompassing the basins of the Tons and the Yamunā rivers, it is admitted that no data is available. But the palaeolakes in the valleys of the Bhāgirathi (Gangā) and its tributary Bhilanganā provide dependable data from multi-proxy studies. The Bhāgirathi catchment is next to the catchment of the Yamunā–Tons. The pollen- and magnetic-susceptibility profile of an alpine peat in the Dokriani Glacier (not far from the Gangotri) unfold a history of a warm–humid climate with highest Monsoon rains in the period 6000 cal yr B. P. to 4500 cal yr B.P., followed by a time (4000–3500 yr B.P.) when there was progressive cooling and greatest increase of the evergreen oak trees, with decreased Monsoon rain reaching the lowest level at 3500 cal yr B.P. (Phadtare 2000). The 4000–3500 yr B.P. is described as the coldest interval of the Holocene epoch.

The basin of Badanitāl in the Bhilanganā valley, next to Bhāgirathi, experienced cold dry climate in the period 5100 yr B.P. to 3500 yr B.P., followed by a time of moist–wet condition in the period 3500 yr B.P. to 1800 yr B.P. (Kotlia et al. 2014).

Melt Water Sustained the Saraswatī Flow

It should be obvious from the discussion in the preceding pages that although the Southwest Monsoon (Summer Indian Monsoon) rain declined considerably, the Winter Westerlies (Plate 7.2) remained active even if subdued, and continued to provide precipitation and pile snows in the mountainous catchments of the two branches of the Saraswatī. This is amply borne out by the evergreen forests and vegetations of alpine meadows that flourished in this terrain. If not the monsoonal rain water, the melt water sustained the flow of the Saraswatī in the period the Harappans lived. The Harappans left their abodes not because the rainfall had decreased, but when the river lost all its water due to the capture of its eastern branch by the Yamunā, and by its western branch by the Vipāsh (Beās) River, wholly depriving it of its water discharge.

Chapter 8
Return and Revival of the Saraswatī

The Saraswatī River has returned to quite some parts of the land it watered before and during the Harappan times. But the courses of both the eastern branch (Tamasā/Tons) and the western branch (Shatadru/Satluj) are now quite different not necessarily through the paleochannels. And these two branches have come back in the form of canals dug through the arable and agricultural lands in Haryānā, Punjāb and western Rājasthān (Fig. 8.1). The trunk canals are associated with networks of smaller canals or branches that have brought back not only the greenery but also endowed the land with rich harvests of crops of cereals, cotton and fruits. The new *avatār* of the Saraswatī branches have transformed the environment and the life of the people who had resiliently lived through more than three thousand years and are living today prosperously in the otherwise parched and arid lands. So much water has come through these canals that many a part of the agricultural lands in Haryānā, Panjāb and Rājasthān are woefully waterlogged—alarmingly in some places.

Western Yamunā Canal

The *Western Yamunā Canal* taps the water of the Yamunā at Tājewālā and Hathini Headworks and passes through the districts of Karnāl, Pānipat, and Bahādurgarh to reach Delhi (Fig. 8.1). Another canal, the Āgra Canal, then takes off from the Okhlā Headworks and joins the Bāngangā River (a tributary of the Yamunā)—nearly 32 km downstream of Āgra. The *Hānsi Branch* goes beyond Kanwari and irrigates fields in Jind and Hisār districts. Other branches carry water to agricultural land in Rohtak, Bhiwāni, Jhajjar and Gurgāon districts.

K.S. Valdiya, *Prehistoric River Saraswati, Western India*,
Society of Earth Scientists Series, DOI 10.1007/978-3-319-44224-2_8

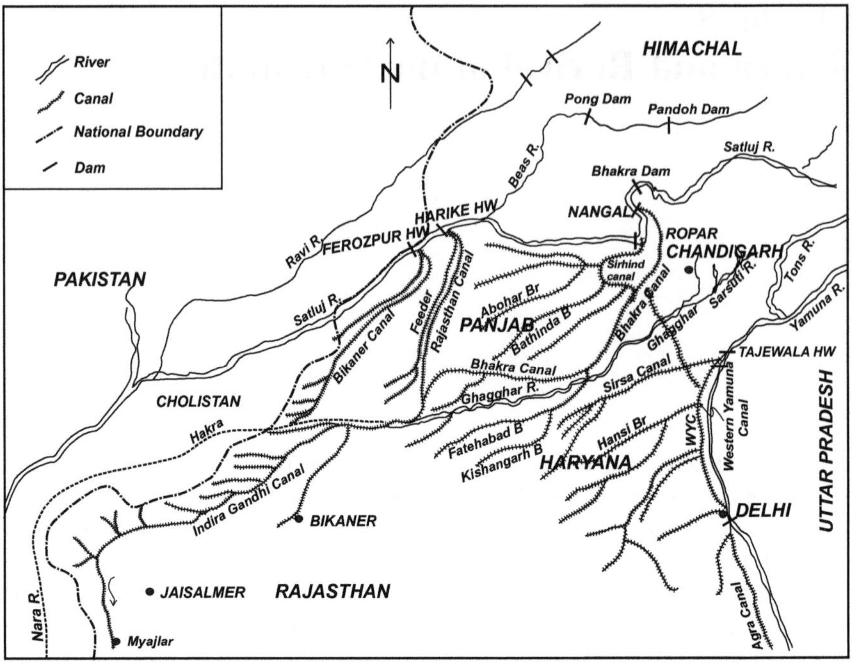

Fig. 8.1 Canals emanating from the once eastern branch of the Saraswatī—the Yamunā (with Tons or Tamasā) and the western branch the Satluj (or Shatadru). Together the two networks of canals irrigate a vast expanse of the Saraswatī land (After Bhatia et al. 2009)

Bhākrā, Sirhind and Rājasthān Canals

The Satluj (Shatadru) water flows through three networks of canals in Panjāb (Fig. 8.1). Originating at Nāngal, the *Bhākhrā Canal* carries the Satluj water through the eastern and southern districts of Panjāb. A major branch of the Bhākhrā Canal flows southeast, linking it with the Sirsā branch of the Western Yamunā Canal. Downstream of the Nāngal Dam is the Ropar Headworks from where the Satluj is tapped by the *Sirhind Canal.* The four branches of this canal irrigate very large part of Panjāb. Its eastern branch closely follows the Ghagghar and then turning west, irrigates the agricultural lands north of the Ghagghar up to the west end of the Sirsā district.

Thanks to the return of the Satluj through the network of canals Panjāb has become the bread basket of India.

Indira Gandhi Canal

The Harike Barrage at Sultānpur in Panjāb diverts a part of the Satluj water, which flows long distance to irrigate the arid country of western Rājasthān—from Kharakhera in Gangānagar district in the north to Myajlār in the Jaisalmer district in the south (Fig. 8.1). The *Bikāner Canal* in western Panjāb joins it in Gangānagar. Over a reach of 167 km the *Indirā Gāndhi Canal* irrigates part of Panjāb and Haryānā and over 445 km in Rājasthān, bringing over 6800 km^2 area under irrigation. The beneficiary districts are Gangānagar, Hanumāngarh, Churu, Bikāner, Jaisalmer and Jodhpur, where crops of mustard, cotton and wheat, among others, are harvested profitably. The course of the Indirā Gāndhi Canal lies quite east of the Hākrā–Nārā reach of the ancient Saraswatī River, but not far from the multitude of abandoned (palaeo) channels of this river in Jaisalmer district which have prolifically yielded sweet water.

Therefore, it would not be wrong to state that the Saraswatī has not abandoned its land wholly. In the form of canals, it is watering the land, and quite excessively in quite many areas.

Revival of the Saraswatī

Efforts to Revive

Concerted efforts at various levels (and involving archaeologists, remote sensing specialists, hydrologists and geologists) to bring back water into the now dry channels of the Saraswatī have been made since 1999 by Saraswatī Nadī Shodh Sansthān, Haryānā under the leadership of Darshan Lal Jain (2015). The Government of Haryānā has been persuaded to undertake deepening and widening of the Sarsuti water courses (including palaeochannels) over a stretch of nearly 100 km in the upper reaches. The offshoot of this programme is that quite a few ancient places, now centres of pilgrimage, such as Kapalmochan at Bilaspur, Kapilmuni Sarovar and Chyavan Kund at Kalāyat (district Jind), and Saraswatī Nagar or Mustafābād (in district Yamunā Nagar) have become popular centres of religious bathing.

The Haryānā Government is reported to have announced a project of building a dam across the Somb River in the foothills of the Siwālik (Figs. 2.2 and 5.1), build a large reservoir at Chhalour area to impound monsoon rainwater, and carry that water through a feeder canal past Rānipur to the dugup channel of the Saraswatī from Rohlaherī (Bilāspur area in Yamunānagar district) 50 km downstream Uncha Chandna-Mustafābād area (Grewal and Sharma 2015).

Digging of this channel yielded fresh water at the depth of over 7 feet on April 21, 2015 at Rohlaheri and also below 5 feet in the first week of May 2015 at Mughalwal (*Hindustan Times*, May 16, 2015). This dug-up course happens to be one of the palaeochannels of the Saraswatī—described as *Prāchi Saraswatī* (eastern Saraswatī), in the 9th century A.D. inscriptions of Pratihār King Bhoj (Grewal and Sharma 2015).

In the arid lower reaches in Jaisalmer district of Rājasthān, Oil and Natural Gas Corporation (ONGC), engaged in drilling for oil, has successfully tapped fresh (potable) fossil water in the fluvial aquifers—dating back to the pre-Harappan to Mature Harappan times, in multiple palaeochannels at depth below 30 m. The ONGC has embarked upon a major project of bringing possible underground Saraswatī water to the surface in this extremely parched land of brackish underground water.

Storage Capacity of Palaeochannels

High resolution techniques of interpretation of satellite imagery have demonstrated existence of multitudes of palaeochannels of the Saraswatī in both Haryānā and western Rājasthān. Drilling for exploration, as already stated has established presence in abundance of freshwater in the aquifers of these palaeochannels in Jaisalmer division of western Rājasthān and in Cholistān across the border. In Haryānā sweet water spouts out in a number of palaeochannels (Bhadra et al. 2006) in the midst of the regime of brackish water, as already stated.

The aggregate length of the large number of identified winding-curving palaeochannels would be of the order of hundreds of kilometres. If water is present all through the extent of these underground conduits (palaeochannels), the volume of underground sweet water-reservoir would be enormous. What has been stated in Chap. 5 provides an example. Despite being intensely used for 40–45 years, there was no indication (in 1998) of any decline in the discharges of the tube-wells or of lowering of the water table in the belts of underground palaeochannels in the heart of the Thār desert in Jaisalmer district (Soni et al. 1999). Interestingly, the freshwater in the aquifers at the depth of 30–50 m is dated 1800–5000 years and at 60–250 m level about 6000–22,000 (Nair et al. 1999). Much older water has been recently discovered in Jaisalmer. This seems to point to the connection of these sweet aquifers in the palaeochannels (Kar and Shukla 2000) to a perennial river that once was (Fig. 8.2).

The palaeochannels are part of riverine aquifers composed of coarse sand along with pebbles in the deeper levels and finer sands and silts. The hydraulic conductivity of the fluvial aquifers (channel deposits) is quite high—higher than that of the alluvial aquifers. In the western Gangā plain, for example, the hydraulic conductivity of the palaeochannel deposits (Fig. 8.3) is 30–75.3 m/day in comparison to 13.5–22.3 m/day of the alluvial aquifer (Samadder et al. 2011). It is therefore logical to surmise that the hydraulic conductivity of the palaeochannel sand bodies in the Saraswatī domain would be equally high.

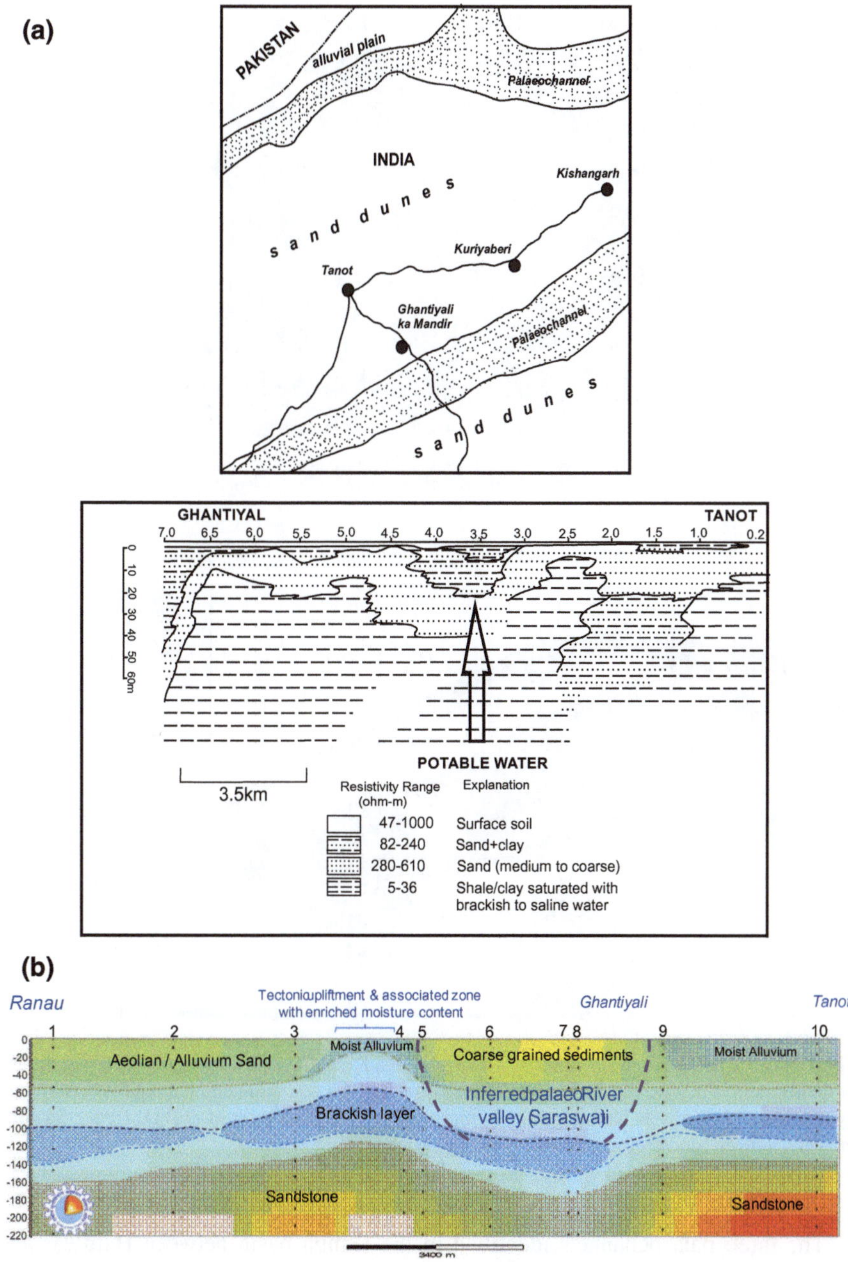

Fig. 8.2 **a** Geophysical investigation for groundwater exploration using geoelectric sounding along a Saraswatī palaeochannel in Jaisalmer district revealed the presence of sweet water in the palaeochannels (After Kar and Shukla 2000). **b** Subsurface electrical section derived from the ground TEM survey conducted along the Ranau–Tanot profile (After Verma et al., 2014; Verma et al., 2016)

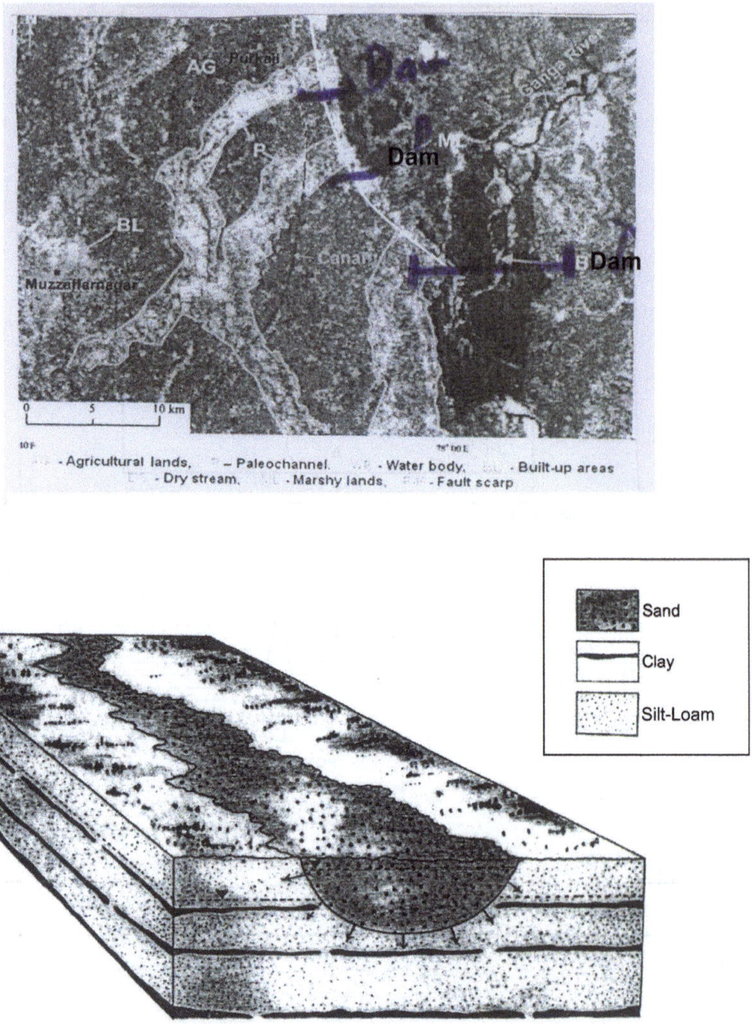

Fig. 8.3 Palaeochannels abandoned by shifting Gangā River have proven to provide great storage of floodwaters. They exhibit higher rate of recharge than the adjoining alluvial aquifers. And their hydraulic conductivity is also appreciably higher. (From Samadder et al. 2011). *Upper* Satellite imagery. *Lower* Model of a segment of the palaeochannel

The three palaeochannels identified in the Gangā basin between Hardwār and Roorkee covering nearly 396 km^2 area have a cumulative water storage capacity of 4.65 billion m^3. One can imagine how much water the multitudes of palaeochannel of the Saraswatī can hold, can store.

'Antahsalilā' Saraswatī

It is evident that an enormous volume of water lies stored—and can be stored—in the palaeochannels, particularly that are 3–4 km wide and 30–60 m deep. Interestingly the underground water in a Gangā palaeochannel NE of Roorkee seeps at the rate of 20–70 m/day (Samaddar et al. 2011). This is the *Antahsalilā*, the underground flowing currents of the river that once flowed on the surface, full of life.

Replenishing Underground Reservoirs

If bore-hole data are integrated with remote-sensing images, the geometry of the underground fluvial aquifer can be reconstructed. By geometry is meant the areal extent, the thickness, the volume, the boundaries and the interconnection with adjacent sand bodies of the aquifers. Admittedly, the geometry will vary from sector to sector, and that tectonic movements such as bulging up, sinking, fissuring and faulting could obliterate the shape and configuration, but the overall pattern of the underground water course and its aquifer would emerge (Fig. 8.3 lower). The hydraulic conductivity being appreciable (20–70 m/day), the water would flow quietly and slowly unseen by us.

The recharge rate of the palaeochannels (Fig. 8.3) being higher (Samadder et al. 2011), it would be rewarding to undertake measures of recharging the multitudes of palaeochannels of the Saraswatī domain wherever feasible in Haryānā (Fig. 1.4). These efforts would sustain the continual flow of the underground currents that flow slowly but surely. This would be tantamount to feeding or recharging the *antah-salila* Saraswatī.

The celebrated waterman of India Rajendra Singh working for over three decades under the banner of Tarun Bhārat Sangh, has revived seven rivers in district Alwar—Arvari, Rupevel Sarsa, Bhagani, Sabi, Jahajwali and Maheshwari in the region of the Arāvali by collecting rainwater and feeding it through fissures and fractures on the river bed. Nearly 11,000 holes were made to recharge the rivers (*The Hindu*, 25 March 2015). These seven rivers have transformed the lives of the people in this part of Rājasthān.

Some Measures of Artificial Recharge

The revival of the Saraswatī River requires implementation of a few measures that would allow surface water to percolate down to the *aquifers of the palaeochannels*. As the first step, major palaeochannels are delineated on the topographic base map by integrating satellite-sensor and well-log data. The CIR composite image of the palaeochannels map are then prepared and, locations of wells indicated and altitudes above sea level shown (Samadder et al. 2011).

Once the total picture is clear, measures are undertaken to feed the crucially located wells, pits, trenches and depressions with storm water flowing through drains. Implied is the action of developing *percolation tanks and infiltration wells* in or on the ground surface of the well-delineated palaeochannels. It is imperative that fine sediments such as clays and silts accumulated in such percolation tanks, wells and ditches are periodically removed to ensure unhindered free percolation and resultant recharge for longer time span.

Spreading floodwaters over the tract of the palaeochannels would be a very effective way of letting water enter into the aquifer, which eventually will find their ways to underground currents.

Water impounded behind dams of all kinds accomplish much the same thing as feeding the percolation tanks and spreading floodwaters do. Therefore, earthen embankments (levees or "*bunds*") *at geologically appropriate locations across and along seasonal streams* and ephemeral rivers, impound water and promote recharge of groundwater effectively. What matters is the water that continually flows underground or in channels on the surface through the land of Haryānā and western Rājasthān.

Water that flows underground escapes evaporation loss due to exposure to hot dry air in the arid–semiarid land. The underground currents can be tapped wherever and whenever needed.

Life-Line of People

The much–adored river known as the Saraswati formed by joining together of the Shatadru (=Satluj) and Tamasa (=Tons) of the past flowed through Haryana, southern Panjab, north-western and western Rajasthan eastern Sindh and emptied itself in the Gulf of Kachchh, as I summarized in my works (Valdiya 2001, 2012, 2013). The Saraswati River was revered by the Rigvedic scholars. It nurtured the Harappa culture until it disappeared in the later Holocene time. The disappearance of the Saraswati River is a case of river piracy by branches of the Ganga and the Sindhu Rivers.

Weaving together various threads of evidence adduced from archaeological, geomorphological and drainage–related studies, together with testimony of sediments laid down by it in its mountain reaches, in its vast floodplain and in the delta, and gleaning relevant information from satellite imageries, it is surmised that the eastern branch of the Saraswati rose in the snowy realm of the Himadri in north-western Uttarakhand. It flowed south-west through one of the tributaries of the present-day Ghaggar River of the foothills and met the western branch, then south-east flowing Shatadru (Satluj), at Shatrana, about 15 km south of Patiala. At the confluence, the channel was 6–8 km wide, implying very high discharge of the Saraswati. The Ghaggar River is known as the Hakra in its middle reaches and as the Nara in the lower reaches. Significantly, the groundwater recovered in the middle reaches from deeper tubewells was found to be 22,000–6000 years old, whereas in the shallow-well the water has been dated 5000–1800 years. The age of the water increased downstream from Kishangarh. Since the tritium value is negligible, these waters do not represent the rainwater fed through contemporary recharge by rainwater. The deeper-and older-water must be attributed to the ancient river that flowed in the time earlier than 5000 yr BP.

Western Rajasthan was dotted with the settlement of the Stone Age people. Parts of Rajasthan, Gujarat and Sindh were inhabited by the people of the Harappa culture (5500–3300 yr BP). More than 2,200 settlements, including those of the Harappa culture and the *ashrams* of sages of the Vedic time, lay on the banks of the River Saraswati that discharged into the Gulf of Kachchh. This is the Saraswati River that the *Rigved* describes in glowing terms—"Breaking through the mountain barrier" this "swift-flowing tempestuous river surpasses in majesty and might all rivers of the land".

Crustal upheavals, such as faulting and attendant displacement of the Siwalik Hills and sinking and rising of the ground in the foothills and attendant the uplift of the NE-SW trending fault-delimited blocks of the Aravali Range, must have caused the deflection of the headwaters of the Yamuna and the Shatadru, leading to the disappearance of this mighty river. The eastern branch deviated southwards around 3700 yr BP, flowed through the channel of a tributary of the Chambal—what is now the Yamuna—and joined the Ganga at Triveni or Allahabad. The consequent dwindling of the river discharge propelled the migration of the Late Harappa (3900–3300 yr BP) people upstream from the Ganganagar-Sirsa area to the upper reaches in the Siwalik. This is evident from a dramatic increase of the Late Harappan settlements in the Siwalik belt in south-eastern Himachal Pradesh, adjoining Haryana and north-western Uttar Pradesh. As a matter of fact, this foothill region became populated for the first time. Later during the time of Gautam Buddha (who lived in the east about 2600 yr BP), the Shatadru River also betrayed the Saraswati River. It abruptly swerved westward to join the Vipash (Beas) of the Sindhu system. Deprived of the waters of these two major rivers, the Saraswati became a dry channel. The collapse of the Harappan culture seems to be wholly due to the disappearance of the Saraswati and associated rivers.

The most scientific, environmentally appropriate and pragmatically feasible way of bringing back to life the lost Saraswati would be to recharge its many paleochannels at many areas by replenishing its undersurface aquifers with floodwaters that can be impounded in critical places identified collectively by remote sensing specialists, geohydrologists, sedimentologists and engineers. The recharged and replenished paleochannels would become the source of water to dug wells and tube wells and to aesthetically developed pools that would attract pilgrims and tourists alike.

Whether flowing underground or on the ground in channels, the Saraswatī has been a life-line of the people from the distant pre-Harappan times some nine thousand years ago to the present. Since the Saraswatī of the yore parted with all its water—a part going to the Gangā through the course of the Yamunā and another part to the Sindhu through the Beas channel—the Saraswatī that now lives comprises the underground currents that flow quietly and very slowly through its multitude of palaeochannels. In another form it flows through the network of canals emanating from the Yamunā and the Satluj rivers.

References

Agrawal A (ed) (2005) In search of Vedic–Harappan relationship. Aryan Books In International, New Delhi

Ahmad F (1986) Geological evidence bearing on the origin of the Rajasthan desert. Proc Ind Natl Sci Acad 52:1285–1306

Allchin R, Allchin B (1993) The birth of Indian civilization. Penguin, New Delhi, p 338

Allchin B, Allchin R (1996) The rise of civilization in India and Pakistan. Cambridge University Press, New Delhi, p 379

Anonymous (2015) Geology and mineral resources of Haryana. In: Miscellaneous Publication 30 (Part XVIII), Geological Survey of India, Kolkata

Auden JB (1935) Traverses in the Himalaya. Rec Geol Surv India 69:123–167

Bakliwal PC, Sharma SB (1980) On the migration of the Yamuna River. J Geol Soc India 21:461–463

Bakliwal PC, Ramasamy SM (1987) Lineament fabric of Rajasthan and Gujarat. Rec Geol Surv India 113:54–56

Bakliwal PC, Grover AK (1988) Signatures and migration of Saraswati River in Thar Desert, Western India. Rec Geol Surv India 116:77–86

Bakliwal PC, Wadhawan SK (2003) Geological evolution of Thar Desert in India—issues and prospects. Proc Indian Natl Sci Acad 69A:151–165

Banerjee S, Chakraborty S (1973) Remains of the Great One-horned *Rhinoceros unicornis* from Rajasthan. Sci Cult 39:430–431

Bansal BK, Verma M (2012) The M 4.9 Delhi earthquake of 5 March 2012. Curr Sci 102:1704–1705

Barnabas S, Suresh CG (1998) Molecular genetic markers in the study of human heritage. In: Balasubramanian D, Appaji Rao N (eds) The Indian Heritage. Universities Press, Hyderabad, pp 27–44

Bhadra BK, Gupta AK, Sharma JR (2006) Subsurface water oozing at Kalayat village, Jind district, Haryana in December 2005: possible connection with Saraswati palaeochannel. J Geol Soc India 68:946–948

Bhadra BK, Gupta AK, Sharma JR (2009) Saraswati Nadi in Haryana and its linkage with the Vedic Saraswati—integrated study based on satellite images and ground-based information. J Geol Soc India 73:273–288

Bhan S (1972) Changes in the course of the Yamuna and their bearing on the Protohistoric cultures of Haryana. In: Deo SB (ed) Archaeological congress and seminar papers, Nagpur, pp 125–128

Bhan S (2000) Aryanization of the Indus civilization. In: Panikkar KN, Byres TJ, Patnaik U (eds) The making of history. Tulika, New Delhi, pp 41–55

Bhatia SB, Singh N (1998) Middle Holocene palaeoclimatic and palaeoenvironmental events in southern Haryana. Proc Indian Natl Sci Acad 54A(4):574–584

Bhatia R, Cestli R, Scaitasta M, Malik RPS (2009) World Bank supported studies: case studies from India, Egypt and Brazil. Academic Foundation, New York

K.S. Valdiya, *Prehistoric River Saraswati, Western India*,
Society of Earth Scientists Series, DOI 10.1007/978-3-319-44224-2

Bhatnagar AK (2014) Astronomical dating from Rigveda to Mahabharat. In: International seminar on Determining Cultural Continuity since Vedic and Epic Eras. Institute of Scientific Research on Vedas, New Delhi

Bisht RS (1978) Banawali: a new Harappan site in Haryana. Man Environ II:86–88

Bisht RS (1982) Excavations at Banawali: 1974–1977. In: Possehl GL (ed) Harappa civilization. Oxford & IBH, New Delhi, pp 113–124

Bisht RS (1984) Structural remains and town planning of Banawali. In: Lal BB, Gupta SP (eds) Frontiers of the Indus civilization. Books & Books, New Delhi, pp 89–97

Bisht RS (1991) Dholavira, new horizon of the Indus Civilization. Puratattva 20:70–82

Bisht RS (1997) Dholavira excavations: 1990–1999. In: Joshi JP (ed) Facets of Indian civilization: recent perspectives, vol I. Aryan Books International, New Delhi, pp 107–120

Bisht RS (1998) Harappans and *Rigveda*: points of convergence. In: Pandey GC (ed) History of science, philosophy and culture in Indian civilization. Books & Books, New Delhi, pp 89–97

Bisht RS (2000) Urban planning at Dholavira: a Harappan city. In: McKim MJ, Gujral LM (eds) Ancient cities—city planning in India. Indira Gandhi National Centre for Arts, New Delhi, pp 16–17

Bisht RS (2013a) Status of Harappan studies as on today—a summary. Institute of Archaeology, Delhi, 24 p

Bisht RS (2013b) Harappan civilization (1921–2013) an overview. Puratattva 43:11–28

Biswas SK (1987) Regional tectonic framework, structure and evolution of western continental marginal basins of India. Tectonophysics 135:307–327

Bohra A, Kotlia BS (2014) Tectonoclimatic signatures during Late Quaternary in the Yunam basin, Baralacha Pass (upper Lahaul Valley), derived from multi-proxy records. Quatern Int. 10.1016/j.quart.2014.10.023

Bose NK, Sen D (1958) The Stone Age in India. The cultural heritage of India. Ramakrishna Mission Institute of Culture, Calcutta, pp 93–109

Bryson RA, Swain AM (1981) Holocene variations of monsoon rainfall in Rajasthan. Quatern Res 16:135–145

Chakrabarti DK, Saini S (2009) The problem of the Saraswati River and notes on the archaeological geography. Aryan Books International, New Delhi

Chakraborti HP (1982) Sarasvati. Great rivers of India. Ramakrishna Mission Ashrama, Patna, pp 9–22

Chatterjeei A, Ray JS (2016) Evidence for a Mid-Holocene buried Himalayan River beneath the Ghaggar Plains: a geochemical provenance study. Goldschmidt Conference Abstracts, NW India, 414 p

Chaudhary P, Bohra NK, Chodhary KR (2011) Conserving biodiversity of community forests and rangelands of a hot arid region of India. Land Use Policy 28:506–513

Chaudhri AR (2008a) Active tectonics in the frontal Siwalik hills in the neighbourhood of Chandigarh, India—a case study. In: International Association for Gondwana Research Conference, Series 5, pp 5

Chaudhri AR (2008b) Civilization along the Saraswati River—a scientific approach. In: Kalyanaraman S (ed) Vedic River Saraswati and Hindu civilization. Aryan Books International, New Delhi, pp 356–377

Clift PD (2009) Harappan collapse. Geoscientist 9(9)

Clift PD, Carter A, Giosan L, Duncan J, Duller GAT, Mcklin MG, Alizai A, Tabrez AR, Danish M, Vanlaningham S, Fuller DO (2012) U-Pb zircon dating evidence for a Pleistocene Saraswati River and capture of the Yamuna River. Geology 40(2):211–214

Courty MA (1995) Late Quaternary environmental changes and natural constraints to ancient landuse (Northwest India). In: Johnson E (ed) Ancient peoples and landscapes. Museum of Technical University, Lubbock, Texas, pp 106–126

Danino M (2010) The Lost River: on the trail of the Saraswati. Penguin, New Delhi, 357 p

de Saint-Martin LV (1860) Elude sur la Geographie et les population, Primitives due northwest de l'Inde, dapres les hymnes Vediques. Imprimeric Imperiable, Paris

Delcaillau B (2006) Recent fold growth and drainage development: the Janauri and Chandigarh anticlines in the Siwalik foothills, NW India. Geomorphology 76:241–256

Deotare BC, Kajal MD, Rajguru SN, Kusumgar S, Jull AJT, Donahue JD (2004) Palaeoenvironmental history of Bap–Balar and Kanod playas of western Rajasthan, Thar Desert. Proc Indian Acad Sci (Earth Planet Sci) 113:403–425

Dhir RP, Singhvi AK (2012) The Thar Desert and its antiquity. Curr Sci 102:1001–1008

Dhir RP, Rajaguru SN, Singhvi AK (1994) Desert Quaternary formations and their morphostratigraphy: implications of the evolutionary history of the Thar. J Geol Soc India 43:435–447

Dimri AP, Chevuturi A (2013) Model sensitivity study for western disturbances over the Himalaya. Meteorol Atmos Phys. doi:10.1007/s00703-013-0302-4

Dixit KN (1981) The excavation at Hulas and further exploration of the upper Ganga–Yamuna. Man Environ V:70–76

Dixit KN (1993) Hulas and the Late Harappan Complex in western Uttar Pradesh. In: Possehl GL (ed) Harappan civilization

Dixit Y, Hodell DA, Petrie A (2014) Abrupt weakening of the summer monsoon in northwest India ~4100 years ago. Geology. doi:10.1130/G35236

Dutta S, Suresh N, Kumar R (2012) Climatically controlled Late Quaternary terrace staircase development in the fold-and-thrust belt of sub-Himalaya. Palaeogeogr Palaeoclimatol Palaeoecol 356–357:16–26

Enzel Y, Ely LL, Mishra S, Ramesh R, Amit R, Lazar B, Rajaguru SN, Baker VR, Sandler A (1999) High resolution Holocene environmental changes in the Thar desert, northwestern India. Science 284:125–126

Flam L (1993) Fluvial geomorphology of the lower Indus basin (Sindh, Pakistan) and the Indus Civilization. In: Shroder JF Jr (ed) Himalaya to the sea. Routledge, London, pp 265–326

Gadgil M, Joshi NB, Shambu Prasad UV, Manoharan S, Patil S (1998) Peopling of India. In: Balasubramanian D, Appaji Rao N (eds) The Indian human heritage. Universities Press, Hyderabad, pp 100–129

Ganguli KM (1998) The Mahabharata of Krishna Dwaipayan Vyasa (12 volumes) Munshiram Manoharlal (Reprints of 1883–1896 editions)

Gaur AS, Vora KH (1999) Ancient shorelines of Gujarat, India during the Indus Civilization (Late-Mid Holocene): a study based on archaeological evidence. Curr Sci 77:180–185

Gaur AS, Vora KH, Sundaresh R, Mani M, Jayakar S (2013) Was the Rann of Kachchh navigable during the Harappan times (Mid Holocene): an archaeological perspective. Curr Sci 105:1485–1491

Geyh MA, Ploethner D (1995) An applied palaeohydrological study of Cholistan, Thar Desert, Pakistan. In: Adar EM, Leibundug C (eds) Application of tracers in arid zone hydrology. International Association of Hydrolocial Sciences, Vienna, pp 119–127

Ghose B (1964) Geomorphological aspects of the formation of salt basins in western Rajasthan. Proceedings, symposium on problems of Indian arid zone. Central Arid Zone Research Institute, Jodhpur, pp 79–83

Ghose B, Kar A, Husain Z (1979) The lost courses of the Saraswati River in the great Indian desert: new evidence from LANDSAT Imagery. Geogr J Lond 145(3):446–451

Ghosh A (1952) The Rajputana desert: its archaeological aspects. Bull Natl Inst Sci India (INSA) 1:37–42

Giosan L, Clift PD, Macklin MG, Fuller DQ, Constaintinescu S, Durcan JA, Stevens T, Duller GAT, Tabrez AR, Gangal K, Adhikari R, Alizai A, Filip FF, Vanlaningham S, Syvitski PM (2012) Fluvial landscapes of the Harappan civilization. In: PNAS. doi:10.1073/pnsas.1112743109 (Early Edition)

Grewal NS, Sharma SK (2015) Unearthing the Saraswati valley. Heritage 1–6

Gupta SP (1996) The Indus-Saraswati civilization. Pratibha Prakashan, Delhi, 110 p

Gupta RK, Prakash I (1975) Environmental analysis of Thar Desert. English Book Depot, Dehradun, 484 p

Gupta AK, Sharma JR, Sreenivasan G, Srivastava KS (2004) New findings on the course of River Saraswati. Photonirvachak 32:1–24

Gupta AK, Bhadra BK, Sharma JR (2008) Saraswati drainage system of Haryana—satellite based study. In: Kalyanaraman S (ed) Vedic Sarasvati and Hindu civilization. Aryan Books International, New Delhi, pp 46–64

Gupta AK, Sharma JR, Sreenivasan G, Srivastava KS (2011a) Using satellite imagery to reveal the course of an extinct river below the Thar Desert in Indo-Pak region. Int J Remote Sens 32 (18):5197–5216 (Published online 1 August 2011)

Gupta RP, Kumar S, Samadder RK (2011b) Palaeochannels as reservoirs for watershed management. In: Proceedings of 4th international conference research in Watersheds, Fairbanks, UK, pp 156–161

Heim A, Gansser A (1939) Central Himalaya: Geological observations of the Swiss Expedition 1936. Denkschr Schweiz Naturf Ges 32:1–245

Iyengar RN (1999) Earthquakes in ancient India. Curr Sci 77:827–829

Iyengar RN (2003) Internal consistency of eclipses and planetary positions in Mahabharat. Indian J Hist Sci 38:75–115

Jain DL (2015) Saraswati project: bringing back the glory of Indian civilization. In: Jain VK (ed) Saraswati Darshan. Saraswati Nadi Shodh Sansthan, Jagadhri, pp 13–14

Jarrige JF, Meadow RD (1980) The antecedents of civilization in the Indus Valley. Sci Am 243 (2):102–110

Jayangondaperumal R, Kumar S, Wesnousky SG, Mahajan AK, Gupta V, Arora BR, Suresh N (2008) Late Pleistocene activity of intrabasinal Bhauwala Thrust, Dehradun, NW Himalaya. Himalayan Geol 29:36

Joshi JP (1972) Exploration in Kutch and excavation at Surkotada, and new light on Harappan migration. J Orient Inst Baroda 22(1):98–144

Joshi JP (1990) Excavation of Surkotada 1971–1972 and exploration of Kutch. Memoirs Archaeological Survey of India, New Delhi

Joshi JP (2008) Harappan architecture and civil engineering. Rupa and Infinity Foundation, New Delhi, 218 p

Joshi JP, Bisht RS (1994) India and the Indus civilization. National Museum Institute, New Delhi

Joshi DD, Dharman R, Saxena AK, Mulkraj (1997) Jaisalmer earthquake of 1991: its effects and tectonic implications. J Geol Soc India 49:433–436

Joshi DD, Kandpal GC, Joshi B, Pande P (2005) Morphotectonic and palaeoseismic studies along the Yamuna tear zone, Himalayan frontal fault belt (Special Publication). Geol Surv India 85:225–236

Juyal N, Pant RK, Yadava MG, Saini NK, Singhvi AK (2004) Climate and seismicity in higher Central Himalaya during 20–10 ka: evidence from the Garbyang basin, Uttarranchal, India. Palaeogeogr Palaeoclimatol Palaeoecol 213:315–330

Kalyanaraman S (2003) Saraswati vol 6 (Language), vol 7 (Epigraphs). Bhabasaheb Apte Smarak Samiti, Bengaluru

Kalyanaraman S (2013) Meluha: a visible language. Saraswati Research Centre, Chennai

Kar A (1988) Possible neotectonic activities in the Luni-Jawar plains, Rajasthan. J Geol Soc India 32:522–526

Kar A (1990) A stream trap hypothesis for the evolution of some saline lakes in the Indian Desert. Z Geomorphol 34:37–47

Kar A (1993a) Aeolian processes and bedforms in the Thar Desert. J Arid Environ 25:83–96

Kar A (1993b) Neotectonic influences on morphological variations along the coastline of Kachchh, India. Geomorphology 8:199–219

Kar A (1996) Morphology and evolution of sand dunes in the Thar Desert as key to sand control measures. Indian J Geomorphol 1(2):177–206

Kar A (1999a) A hitherto unknown palaeodrainage system from the radar imagery of southeastern Thar Desert and its significance. Mem Geol Soc India 42:229–235

Kar A (1999b) Climatic and neotectonic controls on drainage evolution in the Thar Desert. In: Paliwal BS (ed) Geological evolution of northwest India. Scientific Publishers, Jodhpur, pp 246–259

Kar A (2012) GCM-derived future climate of arid, western India and implications for land degradation. Ann Arid Zone 51:147–169

Kar A (2014a) The Thar or the Great Indian Sand Desert. In: Kale VS (ed) Landscapes and landforms of India. Springer Science, Dordrecht, pp 79–90

Kar A (2014b) Agricultural land use in arid western Rajasthan: resource exploration and emerging issues. Agropedology 24:179–196

Kar A, Ghose B (1984) The Drishadvati river system of India: an assessment of new findings. Geogr J 150:221–229

Kar A, Shukla JP (2000) Remote sensing of ground water along the Saraswati palaeochannels in western part of Jaisalmer district, Thar Desert. MAEER's MIT Pune J (Special issue) 1999–2000:28–36

Kar A, Singhvi AK, Juyal N, Rajaguru SN (2004) Late Quaternary aeolian sedimentation history of the Thar Desert. In: Sharma HS, Singh S, De S (eds) Geomorphology and environment. ABC Publications, Kolkata, pp 107–124

Kenoyer JM (1998) Ancient cities of the Indus Valley civilization. Oxford University Press, Karachi

Khadkikar AS, Rajshekhar C, Kumaran KPN (2004) Palaeogeography around the Harappan port of Lothal, Gujarat, western India. Antiquity 78:901

Kochhar R (2008) The Vedic people: Their history and geography. Orient Longman, Hyderabad, 259 p

Kochhar N, Kochhar R, Chakrabarti DK (1999) A new source of primary tin ore in the Indus civilization. South Asian Studies 15:115–118

Kotlia BS, Joshi LM (2013) Late Holocene climate changes in Garhwal Himalaya. Curr Sci 104:911–919

Kotlia BS, Singh AK, Josh LM, Dhaila BS (2014) Precepitation variability in the Indian Central Himalaya during the last ca 4000 years inferred from a speleothem record: impact of Indian Monsoon (ISM) and Westerlies. Quatern Int. doi:10.1016/j.quaint.2014.10.066

Krishna Brahmam N (1993) Gravity and seismicity of the Jaisalmer region, Rajasthan. Curr Sci 64:837–840

Krishna Dvaipāyan Vyās, Before BCE. Purān, *Samvat* 2065. Mahābhārat, Geeta Press, Gorakhpur, 1406 p

Krishnaswamy VS, Jalote SP, Shome SK (1970) Recent crustal movements in northwest Himalaya and the Gangetic fore-deep and related pattern of seismicity. In: Proceedings of 4th symposium earthquake engineering. Roorkee University, Roorkee, pp 419–439

Kshektrimayum KS, Bajpai VN (2011) Establishment of missing stream link between the Markanda river and the Vedic Saraswati river in Haryana, India—geoelectrical resistivity approach. Curr Sci 100:1719–1724

Kumar S, Wesnousky SG, Rockwell TK, Ragona D, Thakur VC, Seitz GG (2001) Earthquakes recurrence and rupture dynamics of Himalayan Frontal Thrust. Science 294:2328–2331

Kumar S, Wesnousky G, Rockwell TK, Briggs RW, Thakur VC, Jayangondaperumal R (2006) Palaeoseismic evidence of great surface-rupture earthquakes along the Indian Himalaya. J Geophys Res 111:Bo3304. doi:10.1029/2004JB003309

Lal BB (1971) Perhaps the earliest ploughed field so far excavated anywhere in the world. Puratattva 4:1

Lal BB (1979) Kalibangan and the Indus Civilization. In: Agrawal DP, Chakrabarti DK (eds) Essays in Indian protohistory. B.R. Publishing Corporation, Delhi, pp 65–97

Lal BB (1997) The earliest civilization of South Asia. Aryan Books International, New Delhi, 287 p

Lal BB (1998) India 1947–1997: New Light on the Indus civilization, New Delhi, pp 116–123

Lal BB (2002) The Saraswati flows on. Aryan Books International, New Delhi, 148 p

Lal BB (2005) The Homeland of the Aryans: evidence of Rigvedic Flora and Fauna. Aryan Books International, New Delhi
Lal BB (2009) How deep are the roots of Indian civilization. Aryan Books International, New Delhi
Lal BB, Thapar BK (1967) Excavations at Kalibangan: New light on Indus civilization. Cultural Forum 9(4):78–88
Lal BB et al (2003) Excavations at Kalibangan, vol I. Archaeological Survey of India, New Delhi
Mackeson F (1844) Report on route from Seersa to Bahawalpore. J Asiatic Soc, XXX, pp 145–150
Malik JN, Mathews G (2005) Evidence of palaeoearthquakes from trench investigation across Pinjore Garden Fault in Pinjore Dun, NW Himalaya. J Earth Syst Sci 114:387–400
Malik JN, Merh SS, Sridhar V (1999) Palaeodelta complex of Vedic Saraswati and other ancient rivers of northwestern India. Mem Geol Soc India 42:163–197
Malik JN, Nakata T, Philip G, Virdi NS (2003) Preliminary observations from a trench near Chandigarh, NW Himalaya and their bearing on active faulting. Curr Sci 85:1793–1798
Malik JN, Nakata T, Philip G, Suresh N, Virdi NS (2008a) Active fault and palaeoseismic investigations: evidence of a historic earthquake along Chandigarh Fault in the Frontal Himalayan Zone, NW India. Himalayan Geol 29:109–117
Malik JN, Morino M, Mishra M, Bhuiyan C, Kaneko F (2008b) First active fault exposure identified along Kachchh Mainland Fault: evidence from trench excavation near Lodai village, Gujarat, northern India. J Geol Soc India 71:201–208
Mallya MR (2005) Ancient India: search for a true chronology. Chaitanya Publication, Bengaluru, 252 p
Marshall J (ed) (1931) Mohenjodaro and the Indus civilization (3 volumes). Arthur Probsthain, London
Mathur UB (2002) Chronology of Harappan port towns of Gujarat in the light of sea-level changes, during the Holocene. Man Environ 27(2):61–67
Maurya DM, Raj R, Chamyal LS (1998) Seismically induced deformational structures (seismites) from the Mid-Late Holocene terraces, lower Mahi Valley, Gujarat. J Geol Soc India 51:755–758
Maurya DM, Khonde N, Das A, Chowksey V, Chamyal LS (2013) Subsurface sediment characteristics of the Great Rann of Kachchh, western India, based on preliminary evaluation of textural analysis of two continuous sediment cores. Curr Sci 104:1071–1077
Merh SS (1993) Neogene-Quaternary sequence in Gujarat: a review. J Geol Soc India 41:259–276
Merh SS (2011) Changing land–sea relationship in the Rann of Kachchh (a geoarchaeological perspective). In: Varadarajan L (ed) Gujarat and the sea. Darshak Itihas Nidhi, Vadodara, pp 11–32
Mishra K (2012) Origin and development of civilization in the Indian subcontinent during the last 8000 years: an archaeological perspective. In: Bala S, Mishra K (eds) Historicity of Vedic and Ramayan Eras. Institute of Scientific Research on Vedas, New Delhi, pp 81–124
Misra VN (1961) Palaeolithic culture of western Rajasthan. Bull Deccan Coll Res Inst 21:86–156
Misra VN (1993) Indus civilization and Rigvedic Saraswati. In: Parpola A, Koskikallio P (eds) South Asian archaeology, vol II., SoumalaineTiedeakatemia, Helsinki, pp 511–525
Misra VN (1995) Lost Saraswati: The Cradle of Harappan civilization. Manthan, XV-4-XVI-1. Deen Dayal Institute, New Delhi, pp 98–116
Mitra DS, Bhadu B (2012) Possible contribution of River Saraswati in groundwater aquifer system in western Rajasthan, India. Curr Sci 102:685–689
Morino M, Malik JN, Mishra P, Bhuiyan C, Kaneko F (2008) Active fault traces along Bhuj Fault and Katrol Hill Fault, and trenching survey at Wandhay, Kachchh, Gujarat, India. J Earth Syst Sci 117:181–188
Mughal MR (1974) New evidence of the Early Harappan culture from Jalilpur, Pakistan. Archaeology 27:106–113
Mughal MR (1984) The post-Harapppan phase in Bahawalpur District, Pakistan. In: Lal BB, Gupta SP (eds) Frontiers of the Indus civilization. Books & Books, New Delhi

Mughal RR (1995) Ancient Cholistan: archaeology and architecture. Ferozsons, Rawalpindi, Karchi, Lahore

Mukherji AB (1976) Terminal fans of inland stream in Sutlej-Yamuna Plain, India. Z Geomorphol NF 20:190–204

Naidu PD (1996) Onset of an arid climate at 3.5 ka in the tropics: evidence from monsoon upwelling record. Curr Sci 71:715–718

Nair AR, Navada SV, Rao SM (1999) Isotope study to investigate the origin and age of groundwater along palaeochannels in Jaisalmer and Ganganagar districts of Rajasthan. Mem Geol Soc India 42:315–319

Nakata T (1972) Geomorphic History of Crustal Movements in the Himalaya. Institute of Geography, Tohuku University, Sendai 77 p

Nath B (1969) The role of animal remains in the early prehistoric cultures of India. Indian Mus Bull

Ngangom M, Thakkar MG, Bhushan R, Juyal N (2012) Continental-marine interaction in the vicinity of the Nara River during the last 1400 years, Great Rann of Kachchh, India. Curr Sci 103:1339–1342

Oldham RD (1886) On the probable change in the geography of the Panjab and its rivers: an historical geographical study. J Asiatic Soc Bengal 55:322–343

Oldham CF (1893) The Saraswati and the lost river of the Indian desert. J R Asiatic Soc Lond 34:49–76

Owen LA, Bailey RM, Rhodes EJ, Mitchell WA, Cox P (1997) Style and timing of glaciations in the Lahul Himalaya, northern India: a framework for reconstructing Late Quaternary palaeoclimatic change in the western Himalaya. J Quatern Sci 12:83–109

Page WD, Aeot JN, Cluff CS, Plafken G (1979) Evidence for recurrence of large magnitude earthquakes along the Makran Coast of Iran and Pakistan. Tectonophysics 52:533–547

Pandey AK, Pandey P (2015) Soft sediment deformation structures in late Quaternary abandoned channel-fill deposit of Yamunā river in NW sub-Himālaya. Curr Sci 108:1717–1725

Pant RK, Juyal N, Basavaiah N, Singhvi AK (2006) Late Quaternary glaciations and seismicity in the Higher Himalaya: evidence from Shalang basin (Goriganga) Uttarakhand. Curr Sci 90:1500–1505

Parkash B, Kumar S, Rao MS, Giri SC, Kumar CS, Gupta S, Srivastava P (2000) Holocene tectonic movements and stress field in western Gangetic plains. Curr Sci 79:438–449

Phadtare NR (2000) Sharp decrease in summer monsoon strength 4000–3500 cal yr B.P. in the Central High Himalaya of India based on pollen evidence from alpine peat. Quatern Res 53:122–129

Philip G, Suresh N, Bhakuni SS, Gupta V (2011) Palaeoseismic investigation along Nalagarh Thrust: Evidence of Late Pleistocene earthquake in Pinjaur Dun, northwestern sub-Himalaya. J Asian Earth Sci 40:1056–1067

Philip G, Bhakuni SS, Suresh N (2012) Late Pleistocene and Holocene large-magnitude earthquakes along Himalayan Frontal Thrust in the Central Seismic Gap in NW Himalaya, Kala Amb., India. Tectonophysics 580:162–177

Philip G, Suresh N, Bhakuni SS (2014) Active tectonism in the northwestern Outer Himalaya: evidence of large-magnitude palaeoearthquake in Pinjaur Dun and Frontal Himalaya. Curr Sci 106:211–222

Pokharia AK, Kharakwal JS, Rawat RS, Osada T, Nautiyal CM, Srivastava A (2011) Archaeobotany and archaeology at Lanmer, a Harappan site in Kachchh, Gujarat: evidence for adaptation in response to climatic variability. Curr Sci 100:1833–1846

Pokharia AK, Kharakwal JS, Srivastava AK (2014) Archaeobotonaical evidence of millets in the Indian subcontinent with some observations on their role in the Indus civilization. J Archaeol Sci 42:442–455

Possehl GL (1999) Indus Age: the beginning. IBH & Oxford Publishing Company, Delhi

Prasad V, Farooqui A, Sharma A, Phartiyal B, Chakraborty S, Bhandari S, Raj R, Singh A (2014) Mid-Late Holocene monsoonal variations from mainland Gujarat, India: a multi-proxy study for evaluating climate-culture relationship. Palaeogeogr Paleoclimatol Paleoecol 397:38–51
Puri VMK (2008) Vedic Saraswati: Scientific signatures on its origin from Himalaya. In: Kalyanaraman S (ed) Vedic Rivier Saraswati and Hindu civilization. Aryan Books International, New Delhi, pp 14–15
Puri VMK, Varma BC (1998) Glaciological and geological source of Vedic Saraswati in the Himalayas. Itihas Darpar 4:7–36
Radhakrishna BP, Merh SS (eds) (1999) Vedic Saraswati: evolutionary History of a Lost River in Northwest India. Geological Society of India, Bangalore, 329 p
Raghav KS (1999) Evolution of drainage basin in parts of northern and western Rajasthan. Mem Geol Soc India 42:175–185
Raikes RL (1964) The end of the ancient cities of the Indus. Am Anthropol 66:284–299
Raikes RL (1969) Kalibangan: death from natural causes. Antiquity XLII
Rajaguru SN, Badam GL (1999) Late Quaternary geomorphology and environment of the Markanda valley, H.P. Mem Geol Soc India 42:143–152
Rajal BS, Virdi NS, Hasija NL (1986) Recent crustal uplift in the Doon Valley. In: Proceedings of international symposium neotectonics in South Asia. Survey of India, Dehradun, pp 146–159
Rajawat AS, Sastry CVS, Narain A (1999) Application of pyramidal processing on high resolution IRS-1C data for tracing migration of the Saraswati River in parts of Thar Desert. Mem Geol Soc India 42:259–272
Rajendran CP, Rajendran K (2000) Geological investigations in the 1819 Kutch (NW India) seismic zone and implications for mechanism of SCR earthquakes. In: Proceedings of international symposium active faults, Hokuden, Japan, pp 387–391
Ramakrishna Rao KV (2008) GVC to IVC to SVC, Ganges Valley Civilization to Indus Valley Civilization to Saraswati Valley Civilization. In: Kalyanaraman (ed) Vedic Sarswati and Hindu civilization. Aryan Books Internatonal, New Delhi, pp 247–274
Ramasamy SM (1999) Neotectonic controls in the migration of Saraswati River in the Great Indian Desert. Mem Geol Soc India 42:153–162
Ramasamy SM, Bakliwal PC, Verma RP (1991) Remote sensing and river migration in western India. Int J Remote Sens 12:2597–2609
Rao SR (1963) Excavations at Rangpur and other explorations in Gujarat. Ancient India 18–19:5–207
Rao SR (1973) Lothal and the Indus civilization. Asian Publishing House, Mumbai
Rao SR (1985) Lothal—a Harappan port town. Mem Archaeol Surv India 78(II)
Rao SR (1991) Dawn and devolution of the Indus civilization. Aditya Prakashan, New Delhi, 392 p
Rao SR (1999) The lost city of Dwaraka. Aditya Prakashan, New Delhi
Rao MR (2015) ONGC Project on Saraswati River. In: Jain VK (ed) Saraswati Darshan. Saraswati Nadhi Shodh Sansthan, Jagadhri, pp 13–14
Rao SM, Kulkarni KM (1997) Isotope hydrology studies on water resources in western Rajasthan. Curr Sci 72:55–61
Rao LS et al (2004/2005) New light on the excavation of Harappan settlement at Bhirrana. Puratattva 35:60–68
Rastogi BK (2014) Seismogenesis of earthquakes occurring in the ancient rift basins of Kachchh, western India. In: Talwani P (ed) Intraplate Earthquakes. Cambridge University Press, New York, pp 126–160
Raverty HG (1892) The Mihran of Sind and its tributaries: geographical and historical studies. J R Asiatic Soc 61(1):155–206 and 297–508.
Rawat S, Gupta AK, Sangode SJ, Srivastava P, Nainwal HC (2015) Late Pleistocene-Holocene vegetation and Indian Summer Monsoon record from the Lahaul, NW Himalaya, India. Quatern Sci Rev 114:167–181

Reddy DV, Nagabhushanam P, Rao MR, Mitra DS, Tripati BM, Sinha SK, Bhadu B (2011) Radiocarbon evidence of palaeo-recharge (pre-Saraswati period, of potential deep aquifer in Thar Desert. J Geol Soc India 77:239–242

Roy AB (1999) Evolution of saline lakes in Rajasthan. Curr Sci 76:290–295

Roy AB, Jakhar SR (2001) Late Quaternary drainage disorganization and migration and extinction of the Vedic Saraswati. Curr Sci 81:1188–1195

Roy B, Merh SS (1977) The Great Rann of Kutch: an intriguing Quaternary terrain. In: Recent advances in geology, vol 9. Hindustan Publishing Corporation, New Delhi, pp 100–108

Rubi BS, Punj NK (1999) Quaternary geological studies in Amritsar district, Punjab. Gondwana Geol Mag 4:33–42

Sahai B (1999) Unravelling of the lost Vedic Saraswati. Mem Geol Soc India 42:121–141

Saini HS, Tandon SK, Mujtaba SAT, Pant NC, Khorana RK (2009) Reconstruction of buried-floodplain systems of the northwestern Haryana Plain and their relation to the 'Vedic' Saraswati. Curr Sci 97:1634–1643

Saini HS, Mujtaba SA (2011) Depositional history and palaeoclimatic variation at northern fringe of Thar Desert, Haryana plain. Quatern Int. doi:10.1016/jgint.2011.06.002

Samadder RK, Gupta RP, Kumar S (2011) Palaeochannels and their potential for artificial groundwater recharge in the western Ganga plains. J Hydrol (14th Feb 2011)

Sanwal J, Kotlia BS, Rajendran CP, Ahmad SM, Rajendran K, Sandiford M (2013) Climatic variability in Central Indian Himalaya during the last ~ 1800 years: evidence from high resolution speleothem record. Quatern Int 304:183–192

Sarkar A, Mukharjee AD, Bera MK, Das B, Juyal M, Morthekai P, Deshpande RD, Shinde VS and Rao LS (2016) Oxygen isotope in archaeological bioapatites from India: implications to climate change and decline of Bronze Age Harappan civilization. Nature Sci Rep. doi:10.1038/srep26555

Sen D, Sen S (1983) Post-Neogene tectonism along the Aravali range, Rajasthan. Tectonophysics 93:75–98

Sharma RS (1999) Advent of the Aryans in India, Manohar Publications, New Delhi

Sharma JR, Bhadra BK (2012) Signatures of palaeorivers network in northwest India using satellite remote sensing. In: Bala S, Mishra K (eds) Historicity of Vedic and Ramayan Eras. Institute of Scientific Research on Vedas, New Delhi, pp 171–192

Sharma JR, Gupta AK, Bhadra BK (2006) Course of Vedic River Saraswati as deciphered from latest satellite data. Puratattva 36:187–195

Shinde VS (1998) Pre-Harappan Padri culture in Saurashtra: the recent discovery. S Asian Stud 13:145–147

Shinde V (2008) Exploration in the Ghagghar basin and excavations at Girawad, Farmana (Rohtak district) and Mitathal (Bhiwani district), Haryana. In: Osada T, Akinori U (eds) Linguistics, archaeology and the human past. Research Institute for Humanity and Nature, Kyoto, p 82

Shroder JF Jr (ed) (1993) Himalayas to the sea. Routledge, London

Shukla MK, Mujtaba SA (2015) Late Quaternary stratigraphy and geomorphic evolution of the cultural plain along the Ghagghar River. In: Extended abstract, seminar on Palaeocene in India. Geological Survey of India, Lucknow, pp 228–229

Shukla AK, Prakash R, Singh RK, Mishra PS, Bhatnagar AK (2007) Seismotectonic implications of Delhi region through fault-plane solution of some recent earthquakes. Curr Sci 93:1848–1953

Siddique SI (1944) River changes in the Ghaggar Plain. Indian Geogr J 19(4):139–146

Singh G (1952) In: The Geography, vol 5, p 27 (cited in: Yashpal et al. 1980)

Singh B (1995) The Vedic Harappans. Aditya Prakashan, New Delhi

Singh NP (1996) Mesozoic–Tertiary biostratigraphy and biogeochronological datum planes in Jaisalmer Basin, Rajasthan. In: Pandey J et al (eds) Contributions 15th Indian Colloquium on Micropalaeontology and Stratigraphy. KDMIPE & WIHG, Dehradun

Singh S (2008) Where lied the Rigvedic rishis, rulers and artisans, the founders of Bharatiaya Sanskriti. In: Kalyanaraman S (ed) Vedic River Saraswati and Hindu civilization. Aryan Books International, New Delhi, pp 118–145

Singh G, Joshi RD, Chopra SK, Singh AB (1974) Late Quaternary history of vegetation and climate of the Rajasthan desert, India. Philos Trans R Soc Lond 267B:467–501

Singh H, Singh R, Raju VS, Chakraborty D (1988) Report on the Quaternary geology and geomorphology of parts of Sirsa and Hissar districts, Haryana. Unpublished Report Geological Survey India, Faridabad

Singh RJ, Joseph DD, Pande P (2008) Geological signature of the Ropar tear in Ropar-Nalagarh sector of the northwestern Himalaya. Himalayan Geol 29:105–108

Singh T, Sharma U, Awasthi AK, Virdi NS, Kumar R (2011) Geomorphic and structural evidences of neotectonic activity in the sub Himalayan belt of Nahan Salient, NW India. J Geol Soc India 77:175–182

Singh A, Paul D, Sinha R, Thomsen KJ, Gupta S (2016) Geochemistry of buried river sediments from Ghaggar Plain: multi-proxy records of variations in provenance, palaeoclimate and palaeovegetation patterns in the Late Quaternary, NW India. Palaeogeogr Palaeoclimat Palaeoecol 449:85–100

Singhvi AK, Kar A (1992) Thar Desert in Rajasthan: land and environment. Geological Society of India, Bangalore, 186 p

Singhvi AK, Kar A (2004) The aeolian sedimentation record of the Thar Desert. Proc Indian Acad Sci (Earth Planet Sci) 113:371–401

Singhvi AK, Kale VS (2009) Paleoclimate studies in India: late ice age to present. Indian National Science Academy, New Delhi, 34 p

Sinha R, Yadav GS, Gupta S, Singh A, Lahiri SK (2012) Geo-electric resistivity evidence for subsurface palaeochannel systems adjacent to Harappan sites in NW India. Quatern Int doi:10.1016/J.Quaint.2012.08.002

Sinha-Roy S (1986) Himalayan collision and indentation of Aravali orogen by Bundelkhand wedge: implications for neotectonics in Rajasthan. In: Proceedings of international symposium neotectonics in South Asia, Survey of India, Dehradun, pp 13–24

Sinha-Roy S, Malhotra G, Mohanty M (1998) Geology of Rajasthan. Geological Society of India, Bangalore, 278 p

Sivewright R (1907) Cutch and the Rann. Geogr J 29:518–535

Snelgrove AK (1979) Migration of the Indus River, Pakistan in response to plate tectonic motions. J Geol Soc India 20:392–403

Soni V (2009) Recharging floodplain. Curr Sci 97:1401–1402

Soni V, Sharma DC, Srivastava KS, Sisodia MS (1999) Hydrogeological, geophysical and isotope study to trace the course of the buried 'Saraswati' River in Jaisalmer district, Rajasthan. In: Paliwal BS (ed) Geological evolution of northwestern India. Scientific Publishers, Jodhpur, pp 305–311

Sood RK, Sahai B (1983) Hydrographic changes in northwestern India. Man Environ 7:166–169

Srivastava GS, Singh IB, Kulshreshtha AK (2006) Late Quaternary geomorphic evolution of Yamuna-Satluj interfluves: significance of the terminal fan. J Indian Soc Remote Sens 34:123–130

Srivastava GS, Kulshreshtha AK, Agarwal KK (2013) Morphometric evidence of neotectonic block movement in Yamuna tear zone of outer Himalaya, India, Zeitsch. Geomorphologie 57:471–484

Srivastava GS, Singh IB, Kulshrestha AK (2013) Surface profiles across Ganga-Indus water divide: Identification of an independent drainage system in Panjab-Haryana plain. Himalayan Geol 34:107–123

Staubwasser M et al (2003) Climate change at 4.2 ka B.P: Termination of the Indus Valley Civilization and Holocene South Asian variability. Geophys Lett 30:1425

Stein MA (1942) A survey of ancient sites along the "lost" Saraswati River. Geogr J 99:173–182

Subramanian TS, Krishnan D (2014) Harappan Surprises, Frontline, June 13, pp 84–98

Talukdar SN, Sudhakar R (1972) Structure of the southern edge of the Himalaya in north India. In: Miscellaneous Publication No. 15. Geological Survey of India, pp 185–194
Tewari HC, Rao VV, Prasad BR (2000) Tectonic significance of seismic reflectivity patterns: a study from the NW Indian Shield. In: Deb B (ed) Crustal evolution and metallogeny in northwestern Indian shield. Narosa, New Delhi, pp 189–202
Thakkar MG, Maurya DM, Raj R, Chamyal LS (1999) Quaternary tectonic history and terrain evolution of the area around Bhuj, Mainland Kachchh. J Geol Soc India 53:601–610
Thakkar MG, Ngangom M, Thakker PS, Juyal N (2012) Terrain response to the 1819 Allahband earthquake in western Great Rann of Kachchh, Gujarat, India. Curr Sci 103:208–212
Thakur VC (1993) Geology of Western Himalaya. Pergamon Press, Oxford, 366 p
Thakur VC (2006) Reassessment of earthquake hazard in the Himalaya and implications from the 2004 Sumatra-Andaman earthquake. Curr Sci 90:1070–1072
Thapar BK (1971) New traits of the Indus Civilization at Kalibangan: An appraisal. S Asian Archaeol 1:85–104
Thapar BK (1975) A Harappan metropolis beyond the Indus Valley. Expedition 17(2):19–32
Thapar R (1979) Ancient Indian social history: some interpretations. Orient Longman, New Delhi, 396 p
Thapar R (2006) The historical beginnings and the concept of the Aryans. National Book Trust, New Delhi
Thusssu JL (1999) Role of tectonics in drainage migration in Punjab-Haryana plain in recent times. Mem Geol Soc India 42:205–217
Thussu JL (2006) Geology of Haryana and Delhi. Geological Society of India, Bangalore, 191 p
Tyagi AK, Shukla AD, Bhushan R, Thakker PS, Thakkar MG, Juyal N (2012) Mid-Holocene sedimentation and landscape evolution in western Great Rann of Kachchh, India. Geomorphology 151–152:89–98
Valdiya KS (1968) Kyā Gangā ne Saraswati ke jal kā apaharan kiyā thā? Dharmayug, 21 July, pp 21–23
Valdiya KS (1980) Geology of kumaun lesser himalaya. Wadia Institute of Himalayan Geology, Dehradun, 291 p
Valdiya KS (1992) The main boundary thrust zone of the Himalaya, India. Ann Tectonicae 6:54–84
Valdiya KS (1996) Saraswati that disappeared. Resonance 1(5):19–28
Valdiya KS (1998) Dynamic himalaya. Universities Press, Hyderabad 178
Valdiya KS (1999a) Rising himalaya: advent and intensification of monsoon. Curr Sci 76:514–524
Valdiya KS (1999b) Marwar's connection with the Himalaya. In: Paliwal BS (ed) Geological evolution of northwestern India. Scientific Publishers, Jodhpur, pp 1–18
Valdiya KS (2001) Reactivation of terrane-defining boundary thrusts in central sector of the Himalaya: implications. Curr Sci 81:1418–1431
Valdiya KS (2002) Saraswati: the river that disappeared. Universities Press, Hyderabad, 116 p
Valdiya KS (2010) The making of India: geodynamic evolution, Macmillan, Noida, 816 p (Chapter 23), p 549–561
Valdiya KS (2012) Geography, peoples and geodynamics of India in Puranas and Epics. Aryan Books International, New Delhi, 240 p
Valdiya KS (2013) The Saraswati was a Himalayan-born river. Curr Sci 104:42–54
Valdiya KS, Pande K (2009) Behaviour of basement-cover decoupling in compressional deformation regime, northern Kumaun (Uttarakhand Himalaya). Proc Indian Natl Sci Acad 75:27–40
Valdiya KS, Joshi DD, Sanwal R, Tandon SK (1984) Geomorphic development across the active Main Boundary Thrust, an example from the Nainital Hills in Kumaun Himalaya. J Geol Soc India 25:761–774
Valdiya KS, Rana RS, Sharma PK, Dey P (1992) Active Himalayan frontal fault, main boundary thrust and Ramgarh Thrust in southern Kumaun. J Geol Soc India 40:509–528

Varma SK, Chandra S, Pavankumar G, Nagaiah E, Ahmed S (2016) Efficacy of time domain electromagnetic method in delineating paleo-channels north of Jaisalmer, Rajasthan, Curr Sci (Communicated)

Van Dijk WM, Densmore AL, Singh A, Gupta S, Sinha R, Mason PJ, Joshi SK, Nayak N, Kumar M, Shekhar S, Kumar D, Rai SP (2015) Linking the morphology of fluvial fan systems to aquifer stratigraphy in the Sutlej-Yamuna plain of northwest India. J Geophys Res Earth Surf 121:1–22. doi:10.1002/2015JF003720

Virdi NS, Philip G, Bhattacharya S (2006) Neotectonic activity in the Markanda and Bata River basin, Himachal Pradesh, NW Himalaya. Int J Remote Sens 27:2093–2099

Wadhawan SK (1990) Quaternary geology, morphostratigraphy and neotectonism in parts of Nagaur district, Rajasthan. Rec Geol Surv India 123(7):16–17

Wadhawan SK (2015) Palaeogene–Neogene tectonics and continental aggradational basins in northwestern India—implications for geological evolution of Thar Desert. In: Abstract national conference palaeogene of the Indian Subcontinent. Geological Survey of India, Lucknow, pp 18–19

Wasson RJ (1995) The Asian monsoon during the Late Quaternary: a test of orbital-forcing and palaeoanalogue forecasting. In: Wadia S, Korisettar R, Kale VS (eds) Quaternary environments and geoarchaeology of India. Geological Society of India, Bangalore, pp 22–35

Wesnousky SG, Kumar S, Mohindra R, Thakur VC (1999) Uplift and convergence along the Himalayan Frontal Thrust. Tectonics 18:967–976

Wheeler REM (1968) The Indus Civilization, 3rd edn. Cambridge Universities Press, Cambridge

Wilhelmy H (1969) Das Urstomtal zum Ostrand der Induse bene und das Saraswati Problem. Z Geomorphol NF Suppl 8:76–93 (The ancient river valley on the eastern border of the Indus plain and the Saraswati problem. Mem Geol Soc India 42:95–112)

Wright RP et al (2008) Water supply and history: Harappa and Beas regional survey. Antiquity 82:37–48

Yash P, Sahai B, Sood RK, Agrawal DP (1980) Remote sensing of the lost Saraswati river. Proc Indian Acad Sci (Earth Planet Sci) 89:317–337

MIX
Papier aus verantwortungsvollen Quellen
Paper from responsible sources
FSC® C105338

If you have any concerns about our products,
you can contact us on
ProductSafety@springernature.com

In case Publisher is established outside the EU,
the EU authorized representative is:
Springer Nature Customer Service Center GmbH
Europaplatz 3, 69115 Heidelberg, Germany

Printed by Libri Plureos GmbH
in Hamburg, Germany